D0908341

DARWIN DELETED

DARWIN
Deleted

Imagining a World without Darwin

PETER J. BOWLER

THE UNIVERSITY OF CHICAGO PRESS
CHICAGO AND LONDON

Peter J. Bowler is professor emeritus of the history of science at Queen's University, Belfast. He has written several books on the development and impact of evolutionism and on science and religion, including *Evolution: The History of an Idea, The Eclipse of Darwinism, The Non-Darwinian Revolution, Charles Darwin: The Man and His Influence, Monkey Trials and Gorilla Sermons, Life's Splendid Drama*, and *Reconciling Science and Religion*, the latter two also published by the University of Chicago Press.

The University of Chicago Press, Chicago 60637
The University of Chicago Press, Ltd., London
© 2013 by The University of Chicago
All rights reserved. Published 2013.
Printed in the United States of America

22 21 20 19 18 17 16 15 14 13   1 2 3 4 5

ISBN-13: 978-0-226-06867-1 (cloth)
ISBN-10: 0-226-06867-6 (cloth)
ISBN-13: 978-0-226-00984-1 (e-book)
ISBN-10: 0-226-00984-x (e-book)

Library of Congress Cataloging-in-Publication Data
Bowler, Peter J., author.
Darwin deleted: imagining a world without Darwin / Peter J. Bowler.
pages cm
Includes bibliographical references and index.
ISBN-13: 978-0-226-06867-1 (cloth: alkaline paper)
ISBN-10: 0-226-06867-6 (cloth: alkaline paper)
ISBN-13: 978-0-226-00984-1 (e-book)
ISBN-10: 0-226-00984-x (e-book)
1. Darwin, Charles, 1809–1882. 2. Evolution (Biology)—History.
3. Imaginary histories.   I. Title.
QH361.B675 2013
576.8'2—dc23   2012033769

♾ This paper meets the requirements of ANSI/NISO Z39.48-1992
(Permanence of Paper).

With apologies to all my friends in the Darwin industry, who would have to find other means of gainful employment in the world envisaged here.

CONTENTS

ILLUSTRATIONS

1

HISTORY, SCIENCE, AND COUNTERFACTUALS

Imagine a dark, stormy night in the South Atlantic at the end of December 1832. Aboard the Royal Navy survey vessel HMS *Beagle* a young naturalist, racked with seasickness, staggers on deck. A sudden wave makes the ship heel violently, and he is washed over the side. The lookout calls "Man overboard!" but it is too dark to see anything in the churning sea, and the storm is too fierce for the officer on watch to risk turning the ship about. Charles Darwin is gone, and Captain Fitzroy will have to face the task of writing to his family in England to break the news. He will certainly tell them that in addition to their personal tragedy, the scientific community has lost a promising young naturalist who might have achieved great things. But he has no idea that Darwin's greatest achievement would have been to write one of the most controversial books of the century, a book that Fitzroy himself would have denounced in public: *On the Origin of Species*.[1]

What would a world without Darwin look like? Many have argued that science would have developed much the same. His theory of evolution by natural selection was "in the air" at the time, an inevitable product of the way people were thinking about themselves and the world they lived in. If Darwin hadn't proposed it, then someone else would have, most obviously the naturalist we know as the "co-discoverer" of natural selection, Alfred Russel Wallace. Events would have unfolded more or less as we know them, although without the iconic term "Darwinism" to denote the evolutionary paradigm. But Wallace's version of the theory was not the same as Darwin's, and he had very different ideas about its implications. And since Wallace conceived his theory in 1858, any equivalent to Darwin's 1859 *Origin of Species* would have appeared years later. There probably would have been an evolutionary movement in the late nineteenth century, but it would have been based on different theoretical foundations—theories that were actually tried out in our own world and that for a time were thought to overshadow Darwin's.

Darwinism was eventually rescued when the new science of genetics undermined the plausibility of the rival theories of evolution following the "rediscovery" of Mendel's laws of heredity in 1900. I suspect that in a world without Darwin, it would have taken until the early twentieth century for the theory of natural selection to come to the attention of most biologists. Evolution would have emerged; science would be composed of roughly the same battery of theories we have today, but the complex would have been assembled in a different way. In our world, evolutionary developmental biology had to challenge the simpleminded gene-centered Darwinism of the 1960s to generate a more sophisticated paradigm. In the non-Darwinian world, the developmental model would have been dominant throughout and would have been modified to accommodate the idea of selection in the mid-twentieth century.

Why is this exercise of any interest at all? If biology ultimately develops toward the same end product, why should anyone care about the possibility that the major discoveries might have been made in an order different from the one we actually experienced? As far as science itself is concerned, the topic may well be academic (in the best sense of the term), but there are wider issues at stake. We might have ended up with similar theories, but we would think about them differently if they had emerged at different times, and this would affect public attitudes toward them.

The impact of Darwin's theory was of course not limited to science itself—it has been seen as a major contributor to the rise of materialism and atheism. Evolutionism offends many religious believers, but of even greater concern is the idea that change is based on chance variations winnowed out by a ruthless struggle for existence. In the eyes of its critics, Darwin's theory of natural selection inspired generations of social thinkers and ideologues to promote harsh policies known as "social Darwinism." Creationists frequently claim that Darwin was directly responsible for generating the vision of Aryan racial superiority that inspired the Nazis to attempt the extermination of the Jews. Apparently it is not enough for critics to challenge Darwinism on allegedly scientific grounds—they contend that it is also immoral and hence dangerous. Even if the scientific evidence is tempt-

ing, one shouldn't consider the theory because it would undermine morality and the social order. But should certain ideas in science be ruled out of
court whatever the evidence suggests?

My interest in exploring what happens in a world without Darwin is
driven by the hope of using history to undermine the claim that the theory
of natural selection inspired the various forms of social Darwinism. The
world in which Darwin did not write the *Origin of Species* would have experienced more or less all of our history's social and cultural developments.
Racism and various ideologies of individual and national struggle would
have flourished just the same and would have drawn their scientific justification from the rival, non-Darwinian ideas of evolution. This is no mere
conjecture, because the real-world opponents of Darwinism were active in
lending support to the ideologies most of us now find so distasteful. Science simply cannot bear the burden imposed on it by those who think it can
inspire whole social movements—on the contrary, science is shaped by the
social matrix within which it is conducted. In the world without Darwin,
the horrors would still exist, but the theory of natural selection would not
have the bogeyman image associated with it by its critics because it would
have been developed too late to play a significant role. We need to think
harder about the wider tensions in our culture responsible for the ideologies that came to have the inoffensive Darwin as their figurehead.

The conjuring of a world in which events followed a different path at
some crucial turning point is known as counterfactual history. It's highly
controversial among historians, although military historians sometimes
like to show how the outcome of a major battle was decided by an event
that seemed trivial at the time but turned out to have momentous consequences. Critics scoff in part because novelists sometimes set their stories in alternate universes, and this underscores the degree of imagination
counterfactual histories require. There are also several schools of historical
thought that assume that the march of events is predetermined by built-in
trends that govern individual action. In these systems there can be no nodal
points at which history could be switched onto a different track. While I
accept that, thanks to broader cultural trends, social Darwinism would have
emerged even without Darwin's theory, I want to explore the possibility
that without Darwin there would not have been a theory of natural selection in the late nineteenth century.

The counterfactual technique faces another level of opposition in the history of science. The scientific method is supposed to offer a foolproof guide to assembling an ever more sophisticated understanding of the real world. That science could have proceeded along paths we did not actually observe might seem to undermine its claim to objective knowledge. If an alternative science is plausible, how can the entities and processes postulated in our theories correspond to the true nature of reality? But we can imagine at least some points in the development of science when there were alternative possibilities of advancement open to researchers, especially if the various routes ended up at the same point later on. To suggest that evolutionism could have emerged without Darwin does not challenge the objectivity of science, although it does invite us to think more carefully about the nature of scientific knowledge.

COUNTERFACTUALS AND HISTORY

Counterfactual history makes sense only if we think that the sequence of events is to some extent open-ended or contingent. There may be some inevitable trends, but there are also nodes from which alternative sequences branch out. In some cases, the turning point is a crucial decision that could be seen at the time as having momentous implications. In others, a fairly trivial event unleashes a train of unanticipated consequences that add up to create a different future.

Ward Moore's 1955 novel, *Bring the Jubilee*, first got me interested in counterfactuals. In the book, a historian from an alternative world in which the Confederates won the battle of Gettysburg and gained their independence invents a time machine to study the battle firsthand. He tries to remain inconspicuous but is spotted by a group of Confederate soldiers advancing toward what will become the battlefield on the following day. They think he is a spy, panic, and turn back. The historian then watches in horror as the battle unfolds along lines that become increasingly unfamiliar to him. Rather than occupying the hills known as the Round Tops, which dominate the battlefield, they allow Union forces get there first and exploit the position to win the battle. The historian is now trapped in a world that will experience a very different sequence of events from those he remembers.

Here is an example of a counterfactual world emerging from an apparently trivial change affecting a few ordinary people, the consequences of which only turn out to be immense when one is in a position to appreciate their cumulative effect. Now consider another scenario, one more familiar to British readers: a world in which the German Luftwaffe won the Battle of Britain in 1940 and the Nazis successfully invaded England. We know that at a crucial point in September 1940 the Royal Air Force (RAF) was reduced almost to impotence because its airfields had been bombed to the point where many were unusable. Then, in a fit of pique after a minor RAF raid on Berlin, Hitler ordered the Luftwaffe to switch its attentions to London. The resulting Blitz destroyed whole areas of the capital city—but the RAF now had time to rebuild its airfields and resume the fight, ultimately defeating the Luftwaffe and, by denying it air superiority, making an invasion impossible. Hitler's decision changed the course of the war: had the assault on the RAF continued, the Germans would certainly have gained control of the air and a successful invasion might have been mounted. In this case the trigger is not a minor event that has unanticipated consequences but a decision made by a key figure that could have been seen at the time as having major implications (even if its full significance was not at first apparent).

The turning point in Darwinian history falls somewhere between these two extremes. Darwin was indeed a key figure without whom the theory of natural selection would not have been developed in anything like the form we know it. But if one imagines him falling overboard on the voyage of the *Beagle*, his death—however tragic at a personal level—would have been perceived as having only minor implications at the time. No one could have suspected that this young naturalist would mature into someone whose ideas would challenge the world. Some events have consequences that are hard to predict and whose significance is not apparent until viewed in hindsight. Most decisions and events get submerged in the general march by forces too strong to be deflected. But counterfactuals depend on identifying nodal points, those rare episodes where it is possible to plead a plausible case that history could have been switched onto a different track.

To make my non-Darwinian universe plausible, I have to defend counterfactual history against critics who claim the technique is fundamentally flawed: history happened just as we know it, and to imagine alternate worlds is pointless. But why is it pointless? Is it because we shouldn't waste time on imaginative fictions, or because the notion of alternate worlds violates what we know about the march of history? The counterfactualist argues for the contingency of history against those who attribute everything to rigid causation or unalterable trends. He or she then has to show that imagining the development of alternate worlds is something more than a parlour game. This can be done by identifying both the triggering events and their consequences, which helps us to grasp the true significance of factors in our own history, factors we all too often take for granted. The novelist constructs an alternate universe to provide an exciting background for a story. But the historian has to show that identifying the nodal points and the alternatives that flow from them helps us to probe the origins of the world we actually live in.

The historian E. H. Carr argued against counterfactuals, insisting that history is a record of what happened and that worrying about might-have-beens is a waste of time. This objection implies a complete lack of interest in historical causation, turning history into a mere record of facts. It also ignores the role of counterfactuals in everyday life—one of the ways in which we learn about the consequences of our actions is to imagine what might have happened had we chosen otherwise. Lawyers too routinely use the counterfactual technique to probe the responsibility of their clients and witnesses. Did the accused realize what the consequences of his or her actions were? One way of testing this is to ask if they considered what might have happened if they didn't take the crucial step. If we can imagine alternative decisions having consequences in everyday life, it seems odd not to extend the possibility to history, which, after all, is the collective product of individual actions. Even philosopher Benedetto Croce, who dismissed the construction of counterfactual worlds as "too wearisome to be long maintained," conceded that we use the technique in our everyday lives and admitted that it was useful to identify which historical events were crucial turning points.[2]

Determinism in History

Croce wanted to defend the role of the individual in history, but most critics of counterfactualism argue that alternate histories are impossible because the course of events is predetermined. There are no nodes at which history could be switched onto a different path because the world is constrained to unfold in a predetermined direction. The direction may be a product of rigid laws of social or cultural evolution, or it may be directed toward an ultimate goal of deep moral significance. Either way, individual decisions can have no effect and there are no actions that can trigger an unpredictable sequence of events. Tolstoy's *War and Peace*, which argues that we cannot blame the French invasion of Russia on Napoleon, offers a classic expression of this view. The French nation was bound to launch an episode of imperial expansion, and if Napoleon had not lived, someone else would have become emperor and would have made the same decisions. Tolstoy's target was the Great Man school of history in which momentous events are triggered by the will of powerfully gifted individuals. I acknowledge the shortcomings of that school of history and have no intention of presenting Darwin as a Great Man who moved the world by sheer willpower. His crucial insight came about because he had a unique combination of interests that allowed him to see links not obvious to others at the time.

The image of the Great Man is associated with the historical writings of Thomas Carlyle, who believed such individuals were sent to transform the world by its Creator. The idea of the Great Man can therefore be understood as irrelevant for counterfactualism because a Great Man is just fulfilling divine will and driving events toward a predestined culmination. He is merely the tool by which historical inevitability imposes its purpose on the world. To make counterfactuals work, leaders such as Napoleon or Hitler have to be able to make idiosyncratic decisions that could not have been foreseen.

Idealists who see history as the unfolding of a divine plan don't have to rely on Great Men to do the job. They often adopt a less hero-centered approach that sees the universe reaching its goal via built-in trends or a predetermined sequence of developmental stages. We are all in our own ways participating in the process, our individual decisions and activities adding up, whether we are aware of it or not, to achieve the next step in the progress toward the final goal. This was the position of Hegel and his followers, and it is reflected in the modern world through the influence of thinkers such as Michael Oakeshott.

Hegel's philosophy of history was turned on its head by Marx—but without losing its determinist implications. E. H. Carr's real objection to counterfactuals was inspired by his Marxism, an ideology he shared with the historian E. P. Thompson. For the Marxist, the laws of social evolution—driven now by economic, not spiritual, forces—ensure the advance of society through a series of states aimed at the ultimate triumph of the proletariat. Idealism and Marxism thus share an antipathy to the possibility that history might be open-ended and unpredictable, although they disagree on the nature of the forces that constrain individual activity within predetermined channels.

Paradoxically, the same sense of a predetermined course of development was inspired by Adam Smith and his fellow economists' sense that human activity was governed by an "invisible hand" ensuring that decisions made by individuals in their own best interests always further the advance of society toward higher levels of efficiency and justice. For as much as they promoted the value of the individual, nineteenth-century liberals used this model to present modern society as the end point of a fixed historical trend. There was a predetermined sequence built into social evolution, ascending from the hunter-gatherer stage though to agricultural feudalism and finally to free-enterprise capitalism. In the hands of anthropologists and archaeologists, this vision of human history provided a model of cosmic evolution in the age of Darwinism. Modern anthropologists still argue that individual actions are determined by the culture within which they are embedded, although they repudiate the idea of predetermined evolution.

All these systems of predestined historical development seem to challenge our sense of free will. How can we make meaningful choices if even great leaders are incapable of making decisions that will alter the predetermined course of events? I am not particularly concerned with the philosophical problem of free will because I think that all historical models can allow for choices in our personal lives. My decisions affect my own life, but the determinist assumes that in the long run individual actions cancel out or are self-correcting so that society as a whole moves in a predictable direction. At best, individuals can only speed up or delay inevitable changes, which is why Marxists become revolutionaries. Even Carr later admitted that if Lenin had lived, the modernization of Russia would have proceeded without the brutalities of the Stalin era. But as a determinist, he held that the economic changes of that time and place were inevitable. Applying this model to the example of the Civil War, if that group of Confederate soldiers had not taken the Round Tops at Gettysburg, another troop would have done the job instead, because the overall pattern of events was fixed.

Contingency and Counterfactuals

Here the counterfactualist steps in and asks, why couldn't a troop from the other side have gotten there first, thus affecting the whole course of the battle? Time travelers aside, the possibility that trivial effect can have major consequences does seem to arise in certain circumstances, especially in the run-up to crucial events, such as battles. The title of Robert Sobel's book about the British defeat in the American War of Independence reminds us of the well-known adage "For the want of a nail, the shoe was lost; for the want of a shoe, the horse was lost; for the want of a horse, the battle was lost . . ."[3]

Defending the role of contingency in history resonates with efforts in evolutionary biology to maintain the open-endedness of the development of life on earth. Stephen Jay Gould famously insisted that if we could go back to the "Cambrian explosion" (when the major animal types first appeared) and rerun the tape of evolutionary history, the outcome might easily have been different and nothing like human beings would ever have appeared. This invokes the very Darwinian point that when we take into account the complex interaction of factors that trigger evolutionary change, different outcomes are often conceivable. To give an obvious example, the possibility of a species invading a new territory may depend on freak meteorological events temporarily allowing geographical barriers to be breached (think of how animals got from South America to the Galapagos Islands). A system in which many independent causal chains interact is always open to what has been called the "butterfly effect" in modern chaos theory—the beat of a butterfly's wing can trigger a chain of events in the atmosphere that eventually produces a hurricane.[4]

In evolutionism, Gould's claim has been challenged by Simon Conway Morris, who argues that the apparent open-endedness of Darwinian evolution is an illusion. There are physical constraints ensuring that the same goals are reached over and over again by different routes—and Morris welcomes the implication of the inevitability of humans appearing on the earth.[5] While examples of evolutionary convergence abound, there are also examples of evolution exploring alternate routes. Think for instance of Australia's lack of placental mammals and the idiosyncrasy of a marsupial world dominated by kangaroos.

The logic of the butterfly effect is a threat not just to the idea of historical trends, but also to old-fashioned materialistic determinism. As the eighteenth-century French scientist Pierre-Simon Laplace observed, if everything is just an assembly of atoms obeying the laws of physics, then an omniscient observer could predict what will happen when even apparently trivial events trigger broader developments. Chaos theory links with many other aspects of modern physics to question whether this old form of determinism is valid. It is no longer clear that we can see the physical universe as a totally predetermined system. Some philosophers and theologians see this gray area as an opportunity to reintroduce the idea of free will, so we are now in a position to accept a role for contingency both through the action of key individuals and through the unexpected consequences of apparently trivial events.

I don't think the usefulness of counterfactuals would be threatened by rigid determinism, as long as open-endedness was preserved in the sense that there is no clearly defined course of development in history. Perhaps the omniscient observer could predict the effect of the butterfly's wing or the firing of the neurons in Hitler's brain when he gave his order to the Luftwaffe in 1940. But—as Croce admitted—recognizing that certain events were key turning points can help us think about the factors involved in any historical or evolutionary outcome. Even if we don't believe that the alternate universe could in fact have emerged, appreciating the fragility of the sequence of events that produced our own world can be useful because it challenges things we take for granted. To pose an effective challenge, of course, the alternatives must possess a certain level of plausibility, and that forces us to confront our assumptions about the inevitably of the way things actually turned out.

Against the exponents of historical determinism there has been a stream of historians willing to use the possibility of alternative universes to probe our understanding of the events that shaped our own world. Winston Churchill, no mean historian when he was out of office, wrote an essay on the possibility of a Confederate victory in the Civil War in a 1932 collection entitled *It Happened Otherwise*. G. M. Trevelyan imagined a world in which Napoleon won at Waterloo (which he might have done had the Prussians arrived a few hours later).[6] More recently Robert Sobel suggested that the British might have won at Saratoga had it not been for problems in getting supplies through to Burgoyne's army. Perhaps the classic use of the counterfactual technique is Robert Fogel's argument that the advent of the railways did not have the crucial effect on American economic development that everyone has assumed. Fogel used economic statistics to show that development might have proceeded just as rapidly had the old transport system based on canals remained in use. Here the detailed analysis of the counterfactual world plays a vital role by undermining confidence in the assumption that railways were vital to progress.[7]

Fogel was less convincing in suggesting reasons why the railways might not have been introduced; canals can transport goods effectively, but railways also offer rapid transportation for people. Most counterfactual claims focus on identifying the key switching point but then pay less attention to the details of how an alternate universe develops. Outlining the main initial difference is one thing, but following that up with a convincing story of how things unfolded thereafter is usually much more difficult. All too often this is an exercise in unbridled imagination, which is why it is usually left to novelists who set exciting stories in an alternate universe. Classics of the genre include Ward Moore's story of a Confederate victory in the Civil War and two accounts set in worlds where Nazi Germany won World War II, Robert Harris's *Fatherland* and Philip K. Dick's *Man in the High Castle*. Closer to my own theme is William Gibson and Bruce Sterling's *The Difference Engine*, a story set in a Victorian Britain ruled by industrialists aided by steam-powered computers. Here again we encounter the theme of technological innovation striking out in a different direction from the one we experienced. Perhaps this similarity is pointing us toward a useful model for understanding counterfactuals, bearing in mind that new technologies have immense effects on our social and cultural development.

The striking point about technological innovation is that it can be extremely competitive. Inventors and industrialists are constantly developing new machines and techniques that have to compete for a place in the market. They are fighting against not only older technologies, but also against rivals who can fulfill a similar role to their own or who could deflect public interest away from their area of application. Such competitive situations cropped up when the builders of steamships strove to replace sail or when electricity companies fought to replace gas as a source of lighting and power. There was an equally fierce battle within the electrical industry between the proponents of direct and alternating current supply.[8]

In all of these cases, the deeper historians investigate, the more contingent the outcome seems. We assume that the developments we actually witnessed were the inevitable outcome of an obviously superior technology displacing a less-efficient rival. But that is often not how it seemed at the time, because the competition was much more finely balanced than we perceive with hindsight. In these circumstances it is easy to imagine how a single event—the death of an inventor or businessman, an accident that generates bad publicity—could affect the outcome. Alterative worlds with different technologies are not as implausible as one might think, because the process of innovation and implementation involves endless competitions, each of which has many possible outcomes.

COUNTERFACTUALS AND SCIENCE

Does what seems plausible for the introduction of new technologies work equally well for scientific discovery? Here we encounter a new problem arising from what is called the realist view of scientific knowledge. If science is building an ever more sophisticated understanding of what nature is really like, how can there be alternative sequences of discovery? When I lecture about the thesis of *Darwin Deleted*, I am frequently accused of lending support to the opponents of science who claim that scientific knowledge is a social construct. The inevitability of scientific discovery is assumed to follow as a consequence of the fact that science creates true knowledge of the world. Since there is only one real world to investigate, there can be only one way to uncover its secrets. To those who see the history of science as a sequence of genuine discoveries about the nature of reality, the claim that there might be alternative ways for science to proceed seems absurd. It implies that theories are human constructs that have no anchor in the real world.

The issue is crucial because in the science wars that have plagued the academic community recently, the validity of scientific knowledge has been hotly contested. Scientists maintain that their critics—postmodernist literary scholars and sociologists of science—seek to undermine the privileged status of scientific knowledge by dismissing its success as owing to mere rhetoric. Far from building up a true picture of the world, scientific theories reflect the shifting sands of intellectual fashion. The scientists, of course, protest and point to the use of their work in the design and construction of the vast number of technological wonders that the modern world takes for granted. If science's pride of place is based on rhetoric, why would anyone feel safe flying in an airplane designed on scientific principles? There may be some postmodernist literary scholars who think scientific texts succeed because they become fashionable, not because they provide information about the real world. But there are few sociologists of science who deny that scientific theories actually *work* as representations of nature. They feel safe when they fly because they know that teams of technical experts can only demonstrate their skills by successfully manipulating the real world. But knowing how to make something work does not guarantee that the theory behind it is a direct blueprint of nature. If we think of theories as models of the world rather than as truths in some absolute sense, it becomes less obvious that there must be a single route by which new discoveries are made.[9]

Theories as Models

Even a realist can admit that the sheer complexity of nature might leave room for alternative strategies for understanding it. No theory can provide a complete description of the way nature works, so other ways of representing a limited section of reality could prove valid. Within each area of scientific inquiry, there may be alternative strategies for pushing research forward. Each alternative would have strengths and limitations, areas where it came close to depicting reality and others where it was less accurate. Contingent circumstances, including the hope of technological spin-off, might determine which route was preferred.

If we soften our commitment to realism, alternative ways of modeling nature become all the more obvious. Theories are the product of human imagination, and this must imply some flexibility in conceptualizing new areas of study. This does not mean that scientists just make things up as they go along, because at each step their model must prove superior to rivals in its ability to explain what is known and predict what will be discovered. Of course, as in the case of competing technologies, it isn't always obvious at first which model is going to succeed. Once one theory starts to gain support, it begins to define which topics are most relevant and which areas of research will be most amenable to investigation, deflecting attention away from its rivals. Observations are, in the technical jargon of the philosophy of science, "theory-laden." Rival theories encourage different methods and techniques of investigation. Several may have the potential to push research forward, and the one that gains the initial advantage has the power to shape the future development of that area of science.

Since the controversy over Thomas Kuhn's *Structure of Scientific Revolutions*, historians have tended to see theories as rival conceptualizations of nature. They have also had to accept the fact that now-rejected models were perfectly capable of promoting apparently valid research, because revolutions in Kuhn's sense are transitions from one functioning research program to another based on a new worldview. All research programs, however successful, eventually run out of steam, and the correlating science enters a crisis state in which innovative thinkers cast around for a new basis upon which to ground further study. At this point Kuhn seems to have thought in terms of rival hypotheses struggling to take over the imagination of the scientific community. Supporters of the determinist view of history must believe that the outcome of this competition is predetermined—only one alternative will allow further advance. The counterfactualist argues that several of the rivals may have the ability to function effectively and contingent circumstances may influence the outcome of the debate.

But surely, the realist-determinist argues, the practical success of dominant theories cannot be an accident. New theories triumph because they offer a more accurate representation of reality and hence pass more experimental tests. A good example of this way of arguing focuses on the case of genetics. The practical success of this science in areas ranging from plant breeding to the latest medical techniques must indicate that there really is something in nature corresponding to the gene. If science had ignored the genetic model, it would have failed to advance. The disaster of the Lysenko affair—a nongenetic theory of heredity adopted in Soviet Russia because it fit the Marxist ideology—shows that the concept of the gene was essential for a true understanding of biology. But historian Greg Radick has pointed out that the success of the gene concept and the failure of Lysenkoism are no longer seen as quite so inevitable. Indeed if one asks, "Is there really something in nature corresponding to the gene?" and defines "gene" in the old-fashioned sense of a chromosomal unit that unambiguously generates a particular characteristic in an organism whatever the environment, then the answer has to be that there is no such thing. That concept, still actively promoted in the popular media, has evaporated at the level of biological research and has been replaced by a number of different concepts of genetic activity, none of which have the same implications.[10]

Viewed in this light, the success of the original oversimplified version of genetics no longer seems quite so inevitable. Radick shows that there was indeed a rival theory available around 1900, and it had the potential to serve as a valid basis for research by focusing on topics that the geneticists ignored. But its originator, the biologist W. F. R. Weldon, died unexpectedly in 1906, leaving a clear path for the geneticists. Here is an obvious example in which a contingent event swung the balance in favor of one theory at the expense of a valid rival, with huge implications for the future development of science.

But was the alternative really plausible at the time? Philosopher Kyle Stanford argues that in principle any theory could be advanced at any time—the problem of unrecognized alternatives.[11] He seems to think that someone in Newton's time could have conceived the principles of relativity, which leaves us worrying that the Newtonians actually adopted an inferior theory. But no historian could accept the claim that such anachronistic ideas could enter scientific discourse. We know that scientists' thinking is constrained both by the knowledge and techniques available to them and by the cultural and social conventions of their time. There was simply no reason for anyone to raise the concerns that shaped Einstein's thinking in the seventeenth century. The question is, are the constraints so rigid that they effectively channel science into only one viable channel of development, or are they loose enough (sometimes, at least) to allow rival concepts to emerge and be tested? The determinist adopts the former position, the counterfactualist adopts the latter.

Plausibility and Counterfactuals

The philosophers and historians who question the viability of counterfactuals are concerned that it is all too easy to imagine rival hypotheses being proposed, but much harder to justify the claim that they could have been taken seriously at the time of their proposal. Any alternative has to be different enough from reality to deflect science onto a new course of development, yet close enough to be acceptable given the pressures exerted by the available facts and the prevailing culture. Many believe that those pressures are so strong that the outcome would be more or less the same whatever the situation of individual scientists. As historian John Henry claims, the drive toward a mechanistic and mathematical model of nature in the seventeenth century was so strong that something like Newtonianism would have emerged even had Newton not been there to supply its foundations.[12]

I am not averse to the suggestion that there are constraints that shape the course of science. My own vision of the world without Darwin assumes that some form of evolutionism would emerge in the late nineteenth century, given the leads provided by scientific discoveries and by cultural developments. It is precisely because there were general trends pushing people toward evolutionism that we can plausibly imagine the general theory of evolution emerging without Darwin. Plausibility is the key problem identified by detractors of the counterfactual approach. All too often any alternative theory turns out to be highly unlikely to have succeeded once the wider situation is taken into account. Perhaps natural selection is a unique example, given that it is hard to think of another case in which a theory was advanced, remained highly controversial for many decades, and only later became accepted by the whole scientific community. In this case we know exactly what the alternatives were, and we know they were viable, because they were widely accepted during the period in which natural selection remained in doubt.

Imagining what would happen without Darwin's theory is worthwhile because, even more than in the case of Fogel's America sans railroad, we have enough evidence from our own world to show that the alternative could work. In this case, at least, the viability of the counterfactual world can be substantiated by hard facts. It will be up to other historians to work out whether Darwin's case is unique. The exercise of imagining a world without his theory will be valuable if it forces us to reexamine links between theories and wider developments that we thought were inescapable. Much of our intellectual baggage may be the product of historical accident rather than the intrinsic conceptual framework of our worldview. Abandoning the assumption that things had to develop in the way they did forces us to think more carefully about why they actually did turn out that way.

Those who assume that the constraints acting on scientists' investigation are so rigid that they completely predetermine the course of development invoke a wide range of factors, empirical and social. There are determinists who think everything is shaped by the logic of scientific discovery, locked into a sequence of revelations about the true nature of the universe. For example, given developments in astronomy and mechanics, someone else would have discovered Newton's laws if Newton himself had not articulated them. Given developments in the study of cells and plant breeding, the concept of the gene would have emerged without the inspiration provided by Mendel. But this positivistic view of science sees as its greatest opponent a position that is equally deterministic from the opposite direction. According to the social constructivists, the course of science is predetermined not by the facts but by the preconceptions of the society within which the scientists operate. Newtonianism was the product of an ideology that modeled the world on the machines that were transforming society. Genetics was biology's response to a society that wanted to breed better plants, animals, and people. But surely the determinist cannot have it both ways. What is determining science—the facts or the ideologies? If one is the real driving force, the other is powerless, and the fact that determinism has two mutually incompatible foundations is probably the best reason for questioning its validity. I am happy to hold the antagonists' coats while they slug it out, because as a believer in contingency, I think both types of influence are valid, but neither is completely prescriptive.

DARWINIAN COUNTERFACTUALS

Darwinism is widely regarded as a prime example of a theory that was bound to emerge when it did. I refer to this as the "in the air" thesis—the idea that the idea of natural selection was a natural expression of the way everyone had begun to think at the time. If Darwin had not been there to articulate it, someone else would have stepped into the breach, and events would have unfolded just the same, except for the lack of the catchy term "Darwinism." Natural selection, however, was by no means an inevitable expression of mid-nineteenth-century thought, and Darwin was unique in having just the right combination of interests to appreciate all of its key components. No one else, certainly not Wallace, could have articulated the idea in the same way and promoted it to the world so effectively.

My attack on the "in the air" thesis draws on the fact that the determinism comes in two mutually contradictory forms. Scientists who favor the positivist or realist version think that the selection theory is a true representation of how nature works, so as soon as the relevant components became available, someone would slot them together in the obvious way. The social determinists claim that Darwinism was a bad theory accepted only because it was an extension of the competitive ideology of Victorian capitalism. They believe that scientists' thinking has been distorted by their tendency to view nature through spectacles tinted with the values of the society in which they live. This position has been endorsed by Marxists and by other left-wing critics of unrestrained free enterprise who see this form of social Darwinism as the real driving force of scientific thinking on the topic. Religious critics offer a somewhat different ideological explanation when they charge that materialism is the true source of scientists' enthusiasm for the selection theory. From T. H. Huxley to Richard Dawkins, the "trial and error" aspect of natural selection has appealed to those who seek to destroy belief in a divinely ordered world.

The contradictions in the social determinists' views on which ideology produced the theory undermine their position. Even if their disagreement could be resolved, the determinists would still be at loggerheads over whether or not Darwinism is a socially induced illusion or—as the scientists claim—a true picture of the world. Regardless, the theory cannot be the "inevitable" product of two entirely different influences. Rather, there must be, at least occasionally, points in the history of science where a new idea can appear unexpectedly and have a significant effect on subsequent developments. If Darwin was unique in the range of experiences he brought to bear on the question, and radical enough to follow an idea that seemed outrageous to most of his contemporaries, then we enter the territory of counterfactual history where it becomes worthwhile to ask just how much difference it would have made if he had not been there to write the *Origin of Species* in 1859.

Extreme Counterfactuals

I am not the first to suggest that Darwin switched the development of science onto a different track, and some of the proposed alternative scenarios are even more radical than the one I shall explore. In *Darwin's Watch*, novelist Terry Pratchett and his scientific collaborators explore some of the implications of the hugely popular Diskworld stories. Here the wizards of the

Unseen University have to make sure that Darwin gets aboard HMS *Beagle*, because if he doesn't, he becomes a country vicar and writes a book entitled *The Theology of Species*. This proposes that evolution is just the unfolding of God's handiwork, and it checkmates the search for a natural explanation, thereby slowing down the advance of science so that humanity becomes extinct because it can't face the challenge of the next ice age. Pratchett recognizes that the *Origin of Species* was indeed a turning point, and his counterfactual history does have the merit of seeing Darwin as the key figure in the emergence of the selection theory. But his alternative universe lacks plausibility because in our own world there were many efforts to promote the idea that evolution is designed by God. But they enjoyed only limited success, and naturalistic theories soon overtook them. The plural "theories" is crucial here, because natural selection was not the only source of the drive toward a more materialistic worldview. Many biologists promoted alternatives that they thought offered better explanations than natural selection, leading to an episode known as the "eclipse of Darwinism."

There is a slightly less dramatic alternative to the one suggested by Pratchett that still ends up in a world without evolutionism. Recalling his attitude toward the question of the origin of new species just before he read the *Origin*, T. H. Huxley wrote that he was so dissatisfied with both the creationist and the evolutionary positions that he was inclined to say "a plague on both your houses." Although drawn ideologically toward scientific naturalism, he found the explanations of evolution suggested at the time unconvincing. He was excited by Darwin's theory not because he was convinced of its adequacy, but because it showed that it was possible to come up with plausible hypotheses on the topic. Huxley's skepticism about the prospects for a scientific evolutionism allows us to imagine a universe even more radically different from ours than the one I explore. He claimed that most of the naturalists who had thought seriously about the topic shared his frustration—so is it possible that the emergence of a scientific evolutionism could have been completely blocked without the input from Darwin?

If Darwin's initiative really was that crucial to the whole evolutionary project, the few other figures keen to promote the idea—Herbert Spencer in Britain, Ernst Haeckel in Germany—might have struggled in vain to get the scientific community to take their non-Darwinian ideas seriously. In such a world, areas such as classification, comparative anatomy, and embryology would not be illuminated by the search for common ancestors, and palaeontology would have remained a purely descriptive science. There would probably be much less attention paid to these areas because they would not be seen as participating in a scientific revolution. More attention would be paid to areas such as physiology and the newly emerging biochemistry, which have always had the advantage of offering practical applications.

We could be living in a world where the biomedical sciences had advanced health care much more rapidly, but biologists remained uninterested in the historical origins of the human body. A surprising number of medical practitioners aren't interested in that topic even in our world—one can fix the human body without knowing how it originated, which is why there are many creationists within the medical profession. So it is quite easy to imagine an alternative world where the biomedical sciences flourished at the expense of evolutionism. No doubt there are many who wish we were living in this alternative universe, and not merely because of the health benefits we might enjoy.

Here is a spectacular counterfactual history of biology, based not just on different theoretical perspectives but also on different research priorities. Scientists are drawn to areas where they feel they can have an impact, and if evolutionism had not seemed attractive, they would have put their energies elsewhere. But is this alternative universe a plausible counterfactual scenario? The claim that science might have missed evolutionism altogether without Darwin rests on the assumption that the kind of skepticism expressed by Huxley was widespread in the scientific community. In fact, though, Huxley was exaggerating in order to highlight the impact of Darwin's innovation. Naturalists had certainly been reluctant to get involved in the development of an evolutionary theory in the 1850s, but by this time little enthusiasm remained for the idea of miraculous creation. In Germany especially, most biologists suspected that natural causes were at work, even though they suspended judgment on what those forces were. There was growing interest in an evolutionary perspective on the fossil record. By the 1850s efforts were being made to develop ideas about how evolution might work, most obviously in Spencer's enthusiasm for the pre-Darwinian mechanism known as Lamarckism (the inheritance of acquired characteristics).[13] Huxley's skepticism was driven by distrust of Lamarckism, even though he was generally favorable toward Spencer's social philosophy. But in this respect he was unusual—even Darwin allowed some role for Lamarckism. Events in our world after the publication of the *Origin of Species* suggest that many scientists took the Lamarckian alternative seriously until it was discredited by genetics in the early twentieth century. If we imagine a decisive effort to promote Lamarckism in the 1860s of the non-Darwinian world, Huxley might have held back, but there are good reasons to believe that many other scientists would have been inspired by it.

A Non-Darwinian Evolutionism

My argument develops a suggestion made briefly by John Waller that without Darwin, the non-Darwinian alternatives would have had a clear run and might have become central to the establishment of an evolutionary world-view long before anyone else could make a convincing case for natural selection.[14] In our world, Lamarckism and the other nonselectionist mechanisms were promoted largely in response to Darwinism, as alternatives designed to limit the apparent materialism of a theory based on random variation and struggle. These nonselectionist theories actually originated in speculations developed earlier in the century, especially in France and Germany. The powerful case for transmutation mounted in the *Origin of Species* prompted everyone to take the subject seriously and begin to think more constructively about how the process might work. Without the *Origin*, few would have paid much attention to Wallace's ideas (which were in many respects much less radical than Darwin's anyway). Evolutionism would have developed more gradually in the course of the 1860s and '70s, with Lamarckism being explored as the best available explanation of adaptive evolution. Theories in which adaptation was not seen as central to the evolutionary process would have sustained an evolutionary program that did not enquire so deeply into the actual mechanism of change, concentrating instead on reconstructing the overall history of life on earth from fossil and other evidence. Only toward the end of the century, when interest began to focus on the topic of heredity (largely as a result of social concerns), would the fragility of the non-Darwinian ideas be exposed, paving the way for the selection theory to emerge at last.

To those who object that Lamarckism and the other nonselectionist theories are simply wrong and could not have become the foundation for an effective evolutionism, I have two responses. The first is that they are no longer as obviously wrong as they appeared only a couple of decades ago.

Modern evolutionary developmental biology ("evo-devo") has not endorsed Lamarckism as such, but it has revived interest in many of the areas of study that were associated with the nonselectionist theories. My second response is that in the late nineteenth century, the inheritance of acquired characteristics was used as the basis for much valid scientific work. In many areas, one can explore the implications of the idea of divergent evolution whatever the mechanism of local adaptation is presumed to be, Lamarckian or Darwinian. So a good deal of late nineteenth-century science could still go forward in a world where the selection theory was not a serious contender. Even in our own world, many scientists did not agree with Darwin's focus on adaptation as the sole driving force of evolution, a perspective revived by evo-devo. Most evolutionists, especially the Lamarckians, saw the development of the embryo as a model for evolution and studied embryology as an integral component of their effort to understand how new characteristics emerge.

The Copernican revolution offers an interesting parallel. The idea that the earth revolves around the sun was explored in the mid-seventeenth century, even though, by our standards, there was no valid physical theory to explain how the planets moved. There was a theory available before Newton's—Descartes cosmology, in which the planets circulated in a vortex of subtle fluid. Although this turned out to be wrong, it allowed a whole generation of scientists to get on with the job of exploring the implications of Copernicanism. Anyone who still thinks that the non-Darwinian theories are simply wrong can see them as the equivalents of the Cartesian cosmology—effective enough at the time to allow for the exploitation of the basic idea of evolution. But if we recognize that some aspects of the non-Darwinian position have reemerged in modern biology, the situation is even more interesting. The most extreme anti-selectionist theories were certainly wrong and might be regarded as blind alleys along which science was led temporarily. But the underlying perspective was not altogether misguided and has been revived after a period of overenthusiastic support for Darwinism in the mid-twentieth century.

The weakness of simple Lamarckism was exposed when biologists began to focus more attention on the mechanism of heredity. At this point, somewhere around 1900, a theory of what is called "hard" heredity (the transmission of characteristics independent of the effects of the environment) would almost certainly have emerged even without any input from

Darwinism. Wider social pressures made the middle classes increasingly worried about the unrestricted breeding of what they considered to be unfit members of the population. The eugenics movement, which called for restrictions on the breeding of the feebleminded and other targeted groups, became immensely influential. It demanded a theory of hard heredity to undermine the credibility of the reformers who claimed that better conditions would improve the unfit. Individual characteristics had to be seen as rigidly predetermined by heredity so they could be eliminated by restrictions on reproduction.

In our world, genetics eventually provided the requisite theory of hard heredity. Radick may well be right that Mendel's laws were not the only foundation on which such a theory could have been based, although I suspect there would have been a tendency to think in terms of unit characteristics, because many early geneticists were influenced by the non-Darwinian idea of evolution by saltations or sudden jumps. A belief in characteristics created as units would mean an expectation that they breed true as units. In our universe, geneticists focused solely on how units are transmitted, turning their backs on the question of how characteristics are formed in the developing embryo. But in a world where natural selection had not been available, the Lamarckians' focus on individual development would have been much more deeply entrenched. Genetics (or its equivalent) would have been integrated into the study of how the material of heredity shapes the features of the organism. Genetic determinism would never have achieved quite the same level of oversimplification as it did in our world, even after we finally came to understand the nature of DNA.

Developments in the study of heredity would have a dramatic effect on evolution theory dominated by non-Darwinian ideas. Eugenics would focus naturalists' attention (for the first time, in a world without Darwin) on the importance of artificial selection. This would provide the perfect model for a policy of preventing the unfit from breeding (and—when carried to extremes—for eliminating them altogether). As Lamarckism began to seem less plausible, the possibility of a natural form of selection would now emerge as a new explanation of adaptive evolution. Something like the genetical theory of natural selection would be formulated in the 1920s and 1930s, which is when Darwinism revived after its eclipse in our own world. But the selection theory would be grafted onto a much wider appreciation of the role played by individual development in evolution, giving something resembling modern evolutionary developmental biology—but reached by a very different route.

The Wider Issues

The emergence of a non-Darwinian evolutionary paradigm would have major consequences for how the theory's broader implications were perceived. This is where the counterfactual technique has an impact on current debates about the implications of evolutionism. The whole point of counterfactuals is to challenge values and attitudes that rest on the assumption that the way things are is the product of historical inevitability. We assume that science and religion must be in conflict and that the debate over Darwinism is a crucial battleground. We are told that the evils of social Darwinism arose from the theory's impact on the ideology of the late nineteenth and early twentieth century—why else use the term? Exploring the non-Darwinian world will suggest that neither of these assumptions is valid. For religious thinkers in the late nineteenth century, taking Darwin's theory out of the equation would make a significant difference, allowing evolutionism to appear in a much less threatening light. But in the area of social attitudes, we shall see that the identification of Darwinism with harsh and immoral policies is misguided. Those policies would have emerged whatever the scientists proposed, and most of what we call "social Darwinism" could be justified equally well through rival theories of evolution.

It would be foolish to suggest that evolutionism could have emerged without any conflict with religious beliefs. The idea that humans emerged from an animal ancestry undermined the traditional Christian view that humans alone are endowed with spiritual qualities. This challenge arises from any theory of evolution, whatever the proposed mechanism of change. It had already become apparent decades before Darwin published the *Origin* and remained a focus of debate even though he hardly mentioned the topic of human origins in his book. But we know that liberal Christians in the 1850s were already becoming more willing to accept a world governed by law rather than miracle. They saw evolutionism as the unfolding of a divine plan and were willing to compromise on the issue of human origins provided evolution was seen as purposeful and progressive. Darwin's opponents pointed out how difficult it was to see a process driven by random variation and a brutal struggle for existence as a manifestation of divine providence. The liberals turned increasingly to non-Darwinian theories such as Lamarckism, which were much easier to reconcile with the belief that a wise and benevolent Creator had established a morally purposeful evolutionary process.

Natural selection made the issue of evolutionism far more controversial because it presented the theory in its most materialistic form. So without Darwin's theory, the path toward evolutionism would have been much smoother. The radical supporters of scientific naturalism would have been robbed of one of their most potent arguments against the concept of nature as a divine artifice. Their materialistic philosophy would still have stirred antagonism, but evolutionism might have played a much less prominent role in the debate—remember that Huxley and company could draw on other areas of science, including developments in physiology. Liberal Christians would have found it easier to checkmate conservative arguments based on the identification of evolutionism with materialism. In our world, Darwinism became a kind of bogeyman, an image invoked to frighten the faithful by highlighting how easy it was for science to undermine faith. Without that symbol, even conservative religious thinkers would have had less reason to fear the threat posed by the general idea of evolution. By the time the selection theory emerged in the 1920s and '30s, it would have had less impact because evolutionism itself was no longer perceived as a threat. There would be a fundamentalist backlash against modernism in America at the time, but it would have less reason to focus on evolutionism as a symbol of the attitudes it distrusted. The counterfactual history of a world without Darwin thus allows us to think about the assumptions underlying our modern debates and to ask whether the antagonism between evolutionism and religion might be a product of particular historical events rather than an inevitable conflict of irreconcilable positions.

Finally, what of social Darwinism? In their efforts to discredit Darwinism, creationists and exponents of intelligent design routinely charge that the theory is responsible for the appearance of immoral social policies up to and including the Nazis' attempt to exterminate the Jews. However, most of the policies and attitudes that have become identified with Darwinism would have emerged even in a world that was unaware of the theory of natural selection. The use of Darwin's name as a symbol to identify these attitudes is an accident of history, not an indication that the theory actually created those attitudes. Darwinism was used to justify many social policies, but rival non-Darwinian theories would have served equally well. Such theories were, in fact, used for that very purpose in our own world. But these links have been forgotten because Darwin became the figurehead for what was perceived as an assault on traditional values. Any mention of evolution, progress, or struggle is automatically perceived as a reference to Darwin's influence, whether or not there was any direct link to his theory. If we can plausibly imagine how the same attitudes could have found scientific justification in a non-Darwinian world, we can expose the prejudices that have allowed Darwin's critics to present him as the cause of the social evils that have emerged over the last century or more.

Left-wing thinkers have sought to discredit Darwinism by presenting it as an outgrowth of Victorian cutthroat capitalism: social Darwinism was possible because the selection theory was actually modeled on the ideology of competitive individualism. Curiously, the claim that Darwinism is bad science has now been taken up by creationists, many of whom (in America, at least) are passionately committed to the same free-enterprise ideology that the Marxists blamed for the creation of Darwinism! They have redefined social Darwinism, ignoring the original link to unrestrained capitalism and focusing instead on the claim that the theory promotes racism and militarism. Given the huge social forces at work, it is hard to take seriously the claim that a scientific theory could actually generate the attitudes that were expressed by generations of racists from slave owners to the Nazis, or the militarism of both Imperial and Nazi Germany. Most scientists would be amazed to hear that their ideas could affect people's thinking so dramatically—indeed, they often complain how difficult it is to communicate their ideas to the general public.

The real evidence against the creationists' charge is that the sources of the attitudes in question are independent of the rise of Darwinism. Aspects of racism and militarism were in place before Darwin published and the (largely forgotten) history of non-Darwinian evolutionism in our own world shows that these theories could be equally well applied to justify what is all too casually called "social Darwinism." The concept of a racial hierarchy was in place long before evolutionism was called in to explain it, and even non-Darwinian evolutionists could appeal to the "struggle for existence." Whether we define "social Darwinism" as cutthroat capitalism, militarism, racism, or eugenics, it can in all cases be linked just as easily to non-Darwinian biological theories. The counterfactual approach will provide the best possible way of making this point, since it will show just how easily science could have been misused in this direction in a world that did not know the theory of natural selection. Instead of blindly assuming that any reference to progress and the struggle for existence must reflect the impact of Darwin's theory, we shall be forced to think more carefully about the complexity of scientific theorizing and its relationship to the wider world.

This argument is not meant to absolve Darwinian from all responsibility. The theory *was* used to justify all the policies mentioned. And there is a serious argument to be had about whether or not the rhetoric inspired by Darwin's harsh image of nature made it easier for extremists to justify their indifference toward certain individuals or groups. But the blanket assumption that all these injustices and horrors were inspired by Darwinism alone simply cannot be sustained once we realize that his was not the only theory of evolution to emerge in the late nineteenth century. Evolutionism in its most general form was almost certainly an expression of late nineteenth-century cultural developments, and as such it is hardly surprising that it became involved with wider social values. But all versions of evolution theory became involved, not just Darwin's. Natural selection was not the most obvious form in which evolutionism could have emerged, and my imaginary non-Darwinian world will help us understand why it is implausible to see Darwinism alone as the cause of what we mistakenly call social Darwinism. To challenge this thesis, one would have to construct an even more bizarre alternate universe, one in which the absence of one man and his theory would have transformed not just the science of the time, but the whole course of social and cultural history.

2

DARWIN'S ORIGINALITY

A seemingly fatal blow to a counterfactual history of evolutionism is the assumption that the theory of natural selection was somehow "in the air" when the *Origin of Species* was published in 1859. Yet Darwin complained in his *Autobiography* that he found it difficult to get most of his readers to understand the theory.[1] If this was so, it's hard to see how it could be a more or less inevitable product of mid-nineteenth-century thought. On the contrary, Darwin's insight was the product of a unique combination of interests and was significantly out of step with the way most of his contemporaries were thinking about the problem.

The "in the air" thesis assumes that once all the components of the theory had become available, it was inevitable that someone would put them together in the right way. If Darwin hadn't done it, someone else would have stepped in, and we would still have had the selection theory, although it would not have been called "Darwinism." In support of this position, other thinkers who can be identified as potential alternatives include, most obviously, Alfred Russel Wallace, whose 1858 paper is routinely hailed as a classic example of the simultaneous discovery of a theory. Several of the standard histories celebrating the centenary of the *Origin* in 1959, including those by Loren Eiseley, John C. Greene, and Gertrude Himmelfarb, repeat this view. More recently a series of books and articles have sought to knock Darwin off his pedestal by arguing that he wasn't really the discoverer of the theory and that supporters who deliberately obscure the contributions of rival claimants are responsible for his heroic image.[2]

The celebrations surrounding the bicentenary of Darwin's birth in 2009 certainly hailed his willingness to challenge traditional ideas. But by endlessly telling the story of his discovery, they may only have encouraged people to believe that what he observed was available to anyone at the time.

Darwin merely won the race to discover natural selection, just as a celebrity athlete does the same thing as everyone else in the field, only slightly better. And the naysayers were constantly nipping at the heels of orthodox historians with their claims that the idea wasn't really that original after all. It would be nice to think that the celebrations cemented Darwin's originality as a thinker, but they may only have reinforced the view that he was merely slotting together information available to anyone at the time.

My project requires me to challenge the "in the air" thesis by showing that Darwin was a truly original thinker and that the *Origin of Species* was a book that no one else at the time could have written (these are related, but not identical, claims). The general idea of evolution was becoming more popular in the years leading up to 1859, so that even without the *Origin*, there would have been a general conversion to evolutionism in the following decade. But the theory of natural selection, with all the applications that Darwin associated with it, is another matter. The "in the air" thesis depends on the assumption that when all the components became available, it was inevitable that they would soon be put together in the right order. The same assumption is made by those who think that natural selection is the "right answer" to the question posed by the evidence for evolution; claiming a unique status for Darwin might, in their view, threaten the objectivity of science. But both of these positions are open to challenge by the counterfactualist.

Greg Radick identifies the key issue in terms of a question: Was the theory of natural selection independent of its history?[3] In other words, was the theory a piece of scientific knowledge that would emerge as soon as enough evidence had accumulated, whatever the social and cultural environment, or was it a contingent product of a particular set of circumstances? The counterfactualist position demands that we accept as an answer the latter half of Radick's question, but it also requires us to go one step further. If we take the view that origin of the selection theory was inseparable from its cultural matrix, then we escape the idea of it being scientifically preordained, but we are subsequently forced into the arms of those who say it was an inevitable outgrowth of Victorian culture. My position rests on the claim that the scientific logic of the theory was far from evident to Darwin's contemporaries, but it also makes the claim that natural selection was not the only—nor even the most obvious—way of translating the ethos of free-enterprise capitalism into science.

As Darwin pointed out, the *Origin* was "one long argument," and it is a shameful denial of his originality to claim that he was merely assembling the pieces of a jigsaw that had been put in front of him. The components of the theory may have been available, but no one else was in a position to put them all together, let alone make a convincing case for the theory. T. H. Huxley famously wrote that when he read the *Origin*, his reaction was, "How extremely stupid not to have thought of that," but he went on to point out that Darwin had shown how the species question could be tackled by considering topics that no one else had realized might be relevant.[4] The basic concept of natural selection may look obvious enough once it has been pointed out, and the components may have been available to all, but Darwin was the first to realize that by putting those components together *in the right context* a major new initiative could be opened up in natural history. Note that despite his enthusiasm for the selection theory as an alternative to the idea of supernatural design, even Huxley did not think it was a complete solution to the question of the origin of new forms.

Alfred Russel Wallace also conceived a basic idea of natural selection, although we shall see that he understood its implications rather differently. Wallace also missed key elements of the case Darwin presented, most obviously the analogy between artificial and natural selection. And far from this being an instance of simultaneous discovery, he came to the idea twenty years later than Darwin. The case for Wallace starting the revolution on his own if Darwin had not been there begins to unravel as soon as one looks at it carefully. If only for purely practical reasons, Wallace could not have written anything with an authority equivalent to the *Origin of Species* until at least a decade later, by which time other naturalists would have moved things in very different directions.

The "in the air" thesis explains too much, or rather it explains the inevitability of the selection theory at two different levels that turn out to be mutually incompatible. Was the theory inevitable because the march of scientific discovery had put in place all the geological, biogeographical, and other forms of evidence that Darwin (or the co-discoverers) needed? Or because the selection theory reflected the competitive ethos of mid-Victorian capitalism so well that it seemed more or less obvious to everyone at the time? Karl Marx noted this prevailing-culture analogy, and later Marxists have continued to insist that the selection theory is bad science because it imposes an ideologically loaded model onto nature. The problem with this argument is that it is also used by other opponents of the selection theory, including creationists who would feel most uncomfortable admitting that they shared any position with Marxism. The attack would be more convincing if its proponents could agree which ideology—materialism or capitalism—was bamboozling scientists into acceptance of the theory.[5]

The scientific and social levels of causation might work in tandem to explain the general trend toward acceptance of evolutionism in the nineteenth century. The theory could not been formulated until the evidence from the fossil record and other sources had begun to make simple divine creation seem implausible. And there were general trends in Western culture as a whole toward greater enthusiasm for a materialistic viewpoint and for the idea of progress, both key factors helping to provide support for evolutionism in biology. Michael Ruse has argued that evolution theory piggybacked its way to popularity as an extension of the idea of progress.[6] But when it comes to explaining the emergence of the selection theory, the question of whether it was "in the air" creates a tension between the scientific and the cultural forms of inevitability. Science is supposed to be international, so how can a successful theory be the product of an ideology dominant in only one or a few countries?

In fact, historians have long abandoned the view that science works in the same way in all cultures. The development of evolutionism in our own world provides clear evidence of this, especially when we note the very different ways in which Darwinism was received in different countries. Science provided the necessary background information, but the conceptualization necessary to formulate the theory required analogies or models that were more readily available in some cultures than in others. British scientists were more likely to be aware of Malthus's principle of population and the work of animal breeders than were their contemporaries in Germany or France. They were also more likely to think in terms of models based on free-enterprise individualism, especially if they came from the upper middle class, as did Darwin. But if natural selection is—as the Marxists claim—merely a projection of capitalist ideology onto nature, why did Darwin think it was so hard to get his readers to understand his theory, and why did other thinkers such as Herbert Spencer treat the implications of Thomas Malthus so differently up to and even beyond the time they encountered the *Origin of Species*? This anti-Darwinian position assumes that the selection theory is bad science tolerated only by societies obsessed with the ideology of free enterprise. Few would accept that view today, yet we find it hard to believe that not everyone in mid-Victorian Britain shared Darwin's cultural background. And hardly anyone is aware that the selection theory was not the only way of using that background as the model for a scientific theory. Once these complexities are recognized, the possibility that Darwin's work and impact were indeed unique becomes more plausible.

The situation becomes even more complex as we begin to appreciate that—for all its strengths—Darwin's theory may not give us a completely adequate account of life's development on earth. For the historian working in the 1960s and 1970s (when I myself came into the field), the success of Darwinism in science encouraged the belief that all alternative perspectives were blind alleys, uniformly rejected as biologists exposed their lack of empirical support. But it turns out that some of those non-Darwinian perspectives may not have been quite as misleading as the neo-Darwinians assumed. Modern evolutionary developmental biology (evo-devo) has encouraged us to recognize not that Darwinism is wrong, but that it might not be telling us the whole story. And some of its "new" perspectives bear a striking resemblance to themes that were once popular alternatives to the one explored by Darwin. Those themes certainly encouraged biologists to explore some blind alleys, but they also held clues to important insights that Darwin and his followers had to marginalize in order to ensure his theory's dominance. If the more balanced perspective of evo-devo is accepted, it becomes even more crucial to understand how and why Darwin was led to the particular version of evolutionism that bears his name.

THE ESSENTIALS OF DARWINISM, PART I:
THE TREE OF LIFE

Darwin succeeded because he applied his theory both as an explanatory tool and as a rhetorical device. The simplistic efforts of those who deny his originality focus on a very basic definition of natural selection and the modern convention of priority in publication (now invariably in a journal article). But glimpsing the foundations of a new concept isn't enough; to get credit as the real discoverer, you have to be in a position to convince your contemporaries that they must take your concept seriously. By reducing the selection theory to a skeleton consisting of the idea of trial and error operating with random variation, it is possible to find other candidates who published the concept before Darwin. But the program advanced in the *Origin of Species* was far more than a skeleton—indeed, it had to be more if it was to have any serious effect on scientists' thinking. Darwin not only fleshed out the selection theory with details of how it operated, explained through effective analogies with artificial selection, he also embedded it in a comprehensive program for reforming whole swathes of biology. Even before his discovery of the selection mechanism, he had worked out how to transform our understanding of the natural world by seeing species as the end products of a divergent, treelike process driven by the adaptation of populations to their local environments in a geographically diverse and geologically unstable world. He spent the next twenty years exploring both the details of the selection process and the implications of treating it as the only, or at least the main, mechanism of evolution. To understand why the *Origin* was so effective, we have to appreciate how Darwin conceived and articulated the various levels at which his theory operated as an explanatory tool.[7]

As we explore how the Darwinian theory worked, it will become apparent that Darwin himself was the only naturalist who could have addressed all the relevant topics in sufficient detail to force his contemporaries to think again about the question of evolution. The discovery of the theory of natural selection was only one component in his drive to articulate a major theoretical revolution that would transform the worldview of natural history (or biology, as we would call it) at several different levels. It is certainly important to appreciate how he combined insights about variation and selection derived from his work with animal breeders with Malthus's principle of population expansion to create the theory of natural selection. But these innovations were only crucial because they could be articulated within a new worldview that he had begun to formulate from the beginning of his theoretical speculations. Darwin was looking for a mechanism like natural selection because he had already realized that evolution had to be seen as a branching process in which species could either die out or divide to spawn a number of descendant species. Once evolution was seen primarily as a process by which populations adapt to their local environments, the possibility that populations could become divided by geographical barriers immediately led to the model of evolution as a branching tree. Natural

Figure 1 · Portrait of Charles Darwin in 1840 by George Richmond.

selection was merely the answer to the question, how does the process of adaptation actually work?

It may be hard to appreciate today, but the model of the "tree of life" as a way of understanding the relationship between species was something quite original when Darwin came to it in the 1830s, although it was becoming more widely recognized by the time he published. The timing is important, though, because it meant that Darwin had a head start over everyone else in trying to understand the implications of a theory of what we now call "common descent." To show how he was able to do this, we must look in some detail at Darwin's life and career, which may seem out of place in a book intended to explore the consequences of what would have happened if he didn't write the *Origin*. But the whole point of focusing on Darwin's originality is to show that no one else had the kind of career and research opportunities to position them to duplicate all of Darwin's work, certainly not at the level of detail and sophistication he provided. Darwin's notebooks from the late 1830s, when he first conceived his theory, the short sketch he wrote in 1842, and the much fuller essay of 1844, constitute the evidence of how he assembled his theory.[8]

Common Descent

Darwin was the first naturalist to thoroughly explore the implications of the model of evolution as a tree in which the branches diverge as isolated populations adapt to their separate environments. This is also known as the theory of common descent, because it explains similarities between species as owing to their retention of fundamental structures derived from a common ancestor, overlaid with more superficial adaptive modifications developed since they separated. Ernst Mayr called the discovery of this model the first Darwinian revolution, which in turn became the foundation for the second revolution constituted by the discovery of natural selection.[9] Mayr's claim that the idea of common descent was itself a major innovation is an important part of the argument that Darwin was a unique figure who was working with a model that no one else in the 1830s had adopted. Others might conceive the idea of natural selection, but without this broader theoretical program, they didn't know what to do with it.

At the start of the century, J.-B. Lamarck had proposed that there might be natural processes adapting species to changes in their environment, but he saw this process as subordinated to a more powerful progressive urge pushing life steadily up the "chain of being."[10] Darwin was perhaps the first to realize that if adaptation to the local environment was the only mechanism of evolution, there would be major implications for the whole system by which species are classified into groups. It was his work on biogeography during the voyage of HMS *Beagle* (1831–36) that allowed him to see how populations could sometimes become divided by geographical barriers. Evolution isn't a process in which a single species advances in a fixed direction: species routinely become divided so that further developments must be seen as multiple branches diverging in response to the challenge of separate environments. Some branches will split over and over again, while others come to a dead end through extinction. The classic example of this branching is the Galapagos finches, although a considerable amount of mythmaking by later scientists and historians has somewhat exaggerated their impact on the young Darwin. But whatever doubts modern scholars may pour on the story of Darwin's finches, the implications of his studies in biogeography cannot be underestimated. Darwin approached the problem of the origin of species from a direction that hardly anyone else at the time would have chosen—not through the succession of species in the fossil record, but through their succession in geographical space. He had the opportunity to make these geographical studies because he gained the position of gentleman-naturalist aboard the *Beagle*, and (as Terry Pratchett's *Darwin's Watch* reminds us) it was by no means inevitable that he should have had this opportunity to travel around the world. If Captain Robert Fitzroy had turned Darwin down because of the shape of his nose (which, as a believer in the importance of physiognomy as a guide to character, he very nearly did), the history of evolutionism would be very different.

The image of the tree of life appeared already in Darwin's notebooks of the late 1830s and was proposed independently by Wallace in a paper published in 1855. Both realized that it explained why naturalists were able to arrange species into groups within groups, using descent from a common ancestor to explain underlying similarities. Closely related species have diverged recently from a common ancestor, while the ancestry of more distantly related forms must be traced further back down the family tree to find the common point of origin. It was only much later that Darwin realized how natural selection might actually drive species ever further apart (and some have claimed that Wallace's 1855 paper influenced him in this), but the basic notion that related species have split off from a common ancestor was embedded in his thinking from a very early stage.

The idea of common descent now seems so obvious that we find it hard to believe that alternative models could have been proposed to account for the relationships among species. But several proposals available in the 1830s deflected attention away from the model of the branching tree.[11] William Sharpe Macleay's quinary or circular system of classification supposed that every genus contained five species that could be arranged in a circle, each family contained five genera, and so on through the taxonomic

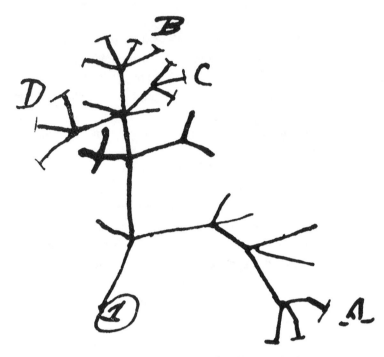

Figure 2 · Darwin's diagram of branching evolution from his B Notebook, 36.

hierarchy. This was a model of nature based on geometrical symmetry. Rob-ert Chambers's hugely influential *Vestiges of the Natural History of Creation* of 1844 discussed Macleay's model but also depicted evolution in terms of parallel lines advancing through a predetermined sequence of stages within each family, each line driven by forces derived from individual develop-ment.[12] Lamarck's followers also tended to focus not on adaptation but on the idea of a linear pattern of progress from simple to complex.

These rigidly structured models of taxonomic relationships and evolu-tion made good sense to anyone wedded to a vision of nature as a predict-able, orderly system governed by a divine plan. Such a worldview made it difficult to accept that the history of life on earth might be essentially irregular and unpredictable, dependant on the hazards of migration, isola-tion, and local adaptation. It was in Germany that this perspective had its most powerful influence, a point that seems to have escaped Darwin, as a result of which it has been almost completely ignored by English-language scientists and historians. Nicolaas Rupke, who is working on a study of the influential German biologist J. F. Blumenbach, calls this the "structuralist" tradition and argues that it led many early nineteenth-century biologists to speculate about natural causes that could produce new forms of life in ways that were predetermined by the laws of nature. This approach included a form of evolutionism, but it offered a very non-Darwinian perspective precisely because it did not employ the model of the tree of life. Instead, it expected to find life unfolding along parallel and predetermined trends in many different parts of the earth.[13]

Adaptation and Biogeography

Why did Darwin move toward a less structured model of how life develops? He was deflected from an interest in rigidly patterned structures because he saw the key problem as that of explaining how species acquire characters allowing them to function effectively in their local environment. He was more interested in adaptation than in cosmic patterns of development, thanks largely to the influence of William Paley's natural theology. Paley demonstrated divine benevolence by pointing to examples of species designed to fit exactly into the environment for which they were created. For Darwin, natural selection replaced divine benevolence as an explanation of adaptation. Unlike Macleay, Chambers, and the Germans, Darwin did not expect his theory to predict an orderly pattern of relationships because he recognized the immense diversity of adaptive challenges to which each species must have responded over the course of its history.

Some scholars argue that Darwin's transition to a more historical viewpoint was inspired by German romanticism.[14] But it seems more probable that his perspective derived from Charles Lyell's geology. There was, of course, one Germanic influence on Darwin, since he had been inspired as a student by Alexander von Humboldt's account of his scientific explorations in South America. Humboldt pointed Darwin toward the importance of biogeography and thus paved the way for the insights he would gain on the *Beagle* voyage. But his pioneering work did not prepare Darwin for the kind of problems that he would encounter when trying to explain the complex relationship between species and geographical barriers. Humboldt had shown that the distribution of species one might expect on the basis of the earth's climatic zones was modified by the uneven distribution of land, sea, and mountains, all of which had major effects on the local environment. But he still thought in terms of clearly defined biogeographical regions, each defined by its characteristic forms of life. What most impressed Darwin on his travels was the crucial role of barriers to migration, past and present, which often seemed to have determined whether or not a species found its way to a particular location. The Galapagos species provided the most obvious example of how the relationships within a group could be explained by supposing that an original population became divided up, in this case by independent acts of migration to oceanic islands. Here Darwin followed his geological mentor Lyell in seeing that biogeography must become a historical science, explaining present distributions in terms of past migrations, extinctions, and (for Darwin but not for Lyell) evolutionary adaptations. He was not the only naturalist to follow Lyell in this respect. Indeed, Edward Forbes anticipated Darwin in publishing an explanation of alpine plants as remnants of an arctic flora that had retreated northward as the last ice age came to an end. But when Forbes moved on to consider the overall distribution of species in time, he came up with a bizarre theory of "polarity" based on an abstract pattern determining the rate of creation in successive geological periods. Forbes thus lost sight of the Lyellian approach and moved toward an idealist model of history based on predetermined trends.[15]

Darwin took the opposite tack, continuing to focus on Lyell's point that we need to explain the existing distribution of species in terms of past migrations. His experiences in South America and the Galapagos convinced him that barriers play a major role in determining how those migrations might take place. As Lyell had emphasized, barriers are not permanent— they can be breached either by changing conditions (such as the diversion of a river's course) or by "accidental" means. Here the Galapagos provided a vital clue, because the islands' populations must have derived from small groups of animals transported from the mainland by lucky accidents. Birds are often blown out to sea by storms, and a few may occasionally make landfall on an isolated island. But the fact that such accidents are unpredictable introduces an element of irregularity into the overall process by which new species come into being. Populations divided by geographical barriers will develop independently as each adapts to its new environment in its own way, and the possibility that barriers can be crossed occasionally allows for the irregularity of the branching process of evolution that Darwin conceived. It is the historical process of migration and local adaptation—in all its unpredictability—that shapes the tree of life. Island biogeography thus played a crucial role in leading Darwin to construct his model of open-ended, divergent evolution. Wallace came to this model by the same route some years later, based on his own experiences in South America and the Malay Archipelago (modern Indonesia).

Darwin's experiences in South America may have pointed him toward the idea of branching evolution in more ways than one. Adrian Desmond and James Moore argue that his hatred of slavery—which he witnessed firsthand in Brazil—may have encouraged him to think in terms of common descent.[16] Since many slaveholders insisted that the black race was created separately from the white, Darwin wanted to show that all the human races share a common ancestry. He then realized that he could defend this claim by extending the idea throughout the animal kingdom. The idea of common descent within humans became a model for the diversity of life as a whole. Desmond and Moore's thesis has generated much controversy, but it emphasizes Darwin's crucial move toward a model of branching evolution based on migration and geographical diversity. It also undermines the widely held assumption that Darwinism and racism are inextricably entwined.

The image of the tree of life was so radical that many late nineteenth-century evolutionists were unable to accept it in full. Ernst Mayr argued that the theory of common descent was one of Darwin's greatest achievements, in addition to natural selection itself.[17] So it was, but I think Mayr overestimated the rapidity with which other naturalists—even those who accepted evolution—were converted to the theory. Many of the non-Darwinian theories of evolution proposed during the eclipse of Darwinism in the late nineteenth century subverted the implications of the principle of common descent.[18] In effect they were a rearguard action to defend the older idea of predetermined development, and in so doing they transferred the Germanic structuralist tradition to British and American science. The fact that such theories continued to flourish demonstrates just how radical the theory of open-ended, divergent evolution was to the naturalists of Darwin's time.

THE ESSENTIALS OF DARWINISM, PART II:
NATURAL SELECTION

Once Darwin had accepted that the adaptation of populations to their local environment was the driving force of evolution, he had to work out how the changes were brought about. One idea was already in circulation: the theory of the inheritance of acquired characteristics. Lamarck had proposed this as a mechanism of transmutation in 1809, although he conceived it as a subsidiary process in addition to progression up the chain of being. Darwin considered adaptation to be the primary factor, and although he never doubted that there was some validity to Lamarck's idea, he did not think it could explain the whole panorama of evolution. In Lamarck's theory, animals change their behavior when challenged by a new environment. In the classic example, when the ancestors of the giraffe found that the grass on which they fed was disappearing, they began to reach up to feed instead from the leaves of trees. As a result of this changed habit, their necks grew longer, just as a weightlifter's arm muscles develop in response to exercise. But Lamarck assumed that these acquired changes could be inherited, something denied by classical genetics but routinely accepted before 1900. The question facing Darwin was, could this effect explain adaptive evolution? He decided it wasn't enough, perhaps because he was suspicious of a theory in which animals could "will" themselves a new structure. He began to look for an alternative and was thus put on the course that led him to natural selection.

To understand how radical his idea was, we need to explain how natural selection works as understood by Darwin and his contemporaries. This involves outlining the components of the process and recognizing the originality of the insights that allowed these elements to be assembled into a theory with such wide-ranging applications. The basic concept of using trial and error as a replacement for intelligent design was outrageously radical for someone working in the 1830s (especially someone from Darwin's social class). But even to recognize the significance of the various components implies a degree of originality, a willingness to think outside the box defined by the conventions of natural history at the time. It was quite unusual to approach the species problem through biogeography, but Darwin's position as gentleman-naturalist on the *Beagle* placed him in a prime position to appreciate the possibilities of this approach. Animal breeding was, of course, widely discussed, but naturalists working on wild species were not in the habit of looking in that direction for inspiration on how to understand natural variation. Here Darwin's position as a country gentleman allowed him immediate access to breeders as a source of ideas and information. Wallace, who came to the same biogeographical insights twenty years later, was in no position to follow up this second line of investigation, even if he had wanted to.

Natural selection is not just a simple application of trial and error, although it does include an element of that principle. It is an iterative process taking place in populations of individuals that reproduce themselves generation after generation, with individual characteristics transmitted to offspring by heredity. But crucially, the copying between generations is not completely accurate, so there is some variation in the sense that new characteristics spontaneously appear from time to time. The selection process operates if some external factor influences the reproductive success achieved by varying individuals. The philosopher Herbert Spencer called it the "survival of the fittest," but the crucial point is that the fittest not only survive more readily than the unfit—they also reproduce more, boosting the proportion of individuals with the fitter characteristic in the next generation. For Darwin (although not quite so clearly for Spencer), fitness was defined solely in terms of an organism's ability to cope with its local environment. There was no absolute standard of fitness, and what was useful in one population might be harmful in another trying to survive in a different environment.

Populations

All of this sounds so obvious that some critics dismiss the idea of natural selection as a tautology—the survival of those who survive.[19] But in fact it is a complex process involving a number of elements, each of which had to be present in the mind of whoever first conceived the theory in a workable form. Most crucially, one has to think in terms of a *population* evolving, and this is so far from being obvious that some historians doubt how completely Darwin himself had converted to what Ernst Mayr called "population thinking." The old idea of the species as a fixed type, with individual variation being merely a trivial deviation from the "true" form (by implication, the form designed by the Creator), has to be abandoned. The species is just the currently breeding population, and if subsequent generations exhibit a different range of characteristics, then the species has by definition evolved. Individual variation is far from trivial; it is an essential feature of the population, and there is nothing to privilege certain characteristics as the "natural" form of the species. Even in the late nineteenth century, there were many naturalists who could not accept this point, preferring theories of evolution by sudden saltations or jumps that produced a new fixed type and a new center for trivial variation.

Mayr championed the view that Darwinism represented the triumph of population thinking over what he dismissed as the old-fashioned typological viewpoint in which each species had an eternally fixed structure.[20] He was, perhaps, too enthusiastic an advocate, and the traditional view of species was by no means committed to a Platonic idealism in which the true design of the species was defined in the mind of God. There were practical reasons why naturalists wanted to see clearly defined species in nature, most obviously because without them there would be little hope of producing a workable system of classification. But Mayr had spotted the important point that anyone who cannot uncouple their thinking from the belief that there is a "true" form on which every species is based will not be able to appreciate the significance of natural selection. At least one of the naturalists who has been hailed as a precursor of Darwin, Edward Blyth, fails on this count. He conceived of selection winnowing out unfit variations, but he saw these as the deviants who had moved too far away from the true form of their species. In his view, selection was a conservative process for protecting the stability of a species, not a mechanism of change.

Darwin certainly began to make the transition to population thinking, even if his writings make some concessions to the traditional view of species. He realized that if a process such as natural selection was at work, then populations that became divided by geographical barriers would begin to diverge from one another until eventually they became new species— but in the intermediate phase, it would be a matter of naturalists' judgment as to whether the diverging populations should be counted as varieties of a single, original species or as the incipient stages of entirely new species. Species thus began to lose their status as clearly defined entities, making it more difficult to sustain the view that they had a permanent essence predetermined by their Creator.

Gradualism

Redefining the nature of species drives home another characteristic of the worldview implied by the selection theory: change must be seen as something gradual and continuous. The rate of change depends on the level of individual variation within a population, and if most of this is fairly trivial, the rate of evolution will be slow because it is defined by the accumulation of many generations of small individual variations. We see very few large-scale individual variants, so the rate of evolution will depend on the amount of ordinary, everyday variation in the population. Theories of saltative evolution tend to assume that rare individuals with a large deviation in form can serve as the foundation for a new species. Mayr and others caricatured such ideas as the theory of the "hopeful monster" and pointed out that in most cases such macromutations would not be able to breed successfully.

Gradualism was fundamental to Darwin's whole way of thinking. The idea that evolution itself is gradual does not, of course, rule out the possibility of major discontinuities caused by external events, as in our modern acceptance of mass extinctions caused by geological or astronomical catastrophes. But Darwin would have none of this. He had originally been trained by Adam Sedgwick, professor of geology at Cambridge, a leading exponent of the "catastrophist" school of thought, which attributed all major changes in the earth's surface to violent upheavals.[21] Sedgwick believed that these events caused mass extinctions, and he would have presumed that some kind of supernatural intervention was required to create new species afterward. On the *Beagle* voyage, however, Darwin was converted to the rival uniformitarian methodology proposed in Charles Lyell's *Principles of Geology*. Lyell distrusted catastrophism because he thought it retained too much of the old scriptural geology inspired by the Genesis account of the creation and Noah's flood. He argued that all the transformations of the earth's surface—deposition of strata, mountain building, erosion, and the like—can be explained by the operation of normal, observable causes over vast periods of time. There was no need to postulate events of a kind that had never been witnessed in the course of human history, and he implied (incorrectly) that the catastrophists thought their great upheavals were caused by miracles. On his travels in South America, Darwin saw evidence of the gradual elevation of the Andes mountain chain by normal earthquakes and became a complete convert to Lyell's point of view.

To understand Darwin's subsequent theorizing, we have to realize how fundamental this conviction was—and how unusual it was at the time. Although Lyell is now hailed as one of the founders of modern geology, at the time there were few converts to his way of thinking. Most geologists reluctantly scaled down the extent of their imagined catastrophes, but very few abandoned them altogether. They were quite right, in one sense, since we now recognize that there have been periods when geological activity was much more violent than it is today (to say nothing of asteroid impacts). Darwin was one of the very few complete converts to uniformitarianism, and without this step in his thinking, he would not have been looking for a mechanism of evolution that was both gradual and firmly based on causes that can be observed in the modern world. Some years later, Wallace too became a convert to Lyell's position, but even he missed out on a key implication that occurred to Darwin. In his search for clues as to what processes might be at work to transform species, Darwin looked to the one area where significant changes can actually be observed: the work of animal breeders and horticulturalists. This supplied him both with evidence about heredity and variation, and also with a crucial analogy that could be used to help his readers visualize natural selection.

An additional factor can be invoked to explain Darwin's enthusiasm for continuity, and more generally for a model treating species as populations of non-identical individuals. The question of an ideological component in his thinking has been controversial ever since Karl Marx noted the parallels between natural selection and the capitalist system based on individualism and free enterprise. Generations of commentators have dismissed the theory as bad science that succeeded only because it gave support to the ideology of laissez-faire. Most scientists insist that the ideological link is superficial and poses no threat to the theory's validity. These two positions can be reconciled, however, at least in terms of how the theory came to be discovered. The realist view that the theory is simply true, and hence stands or falls on the factual evidence, does not rule out the possibility that in order to conceive the idea in the first place, it might be helpful to have a model derived from elsewhere to serve as an inspiration. Natural selection may have been a true theory just waiting to be discovered, but that discovery would have been made more easily in a society that encouraged the particular habits of thought associated with political individualism.

Most historians now accept that it was no accident that natural selection was conceived by British scientists working in the heyday of Victorian capitalism.[22] More specifically, the individualist social model provided by Adam Smith and the political economists of the time might encourage the view that a biological species was just a population of individuals looking after their individual self-interests. The system favored by many Continental thinkers, in which the state is an all-powerful agency directing everyone's activities, provided a closer analogy to the typological view of species. In such a model, change could only come from sudden revolution, not gradual evolution through the summation of individual activities. Darwin came from an upper-middle-class background and was linked to commercially successful families such as the Wedgwoods—just the sort of people who would take naturally to the individualist way of thinking. Social progress was going on all around him, and it was driven by the day-to-day commercial activities of members of his own class. This social context also explains why he would have read Malthus on population and why he would have seized on just those aspects of Malthus's thought that chimed with the individualist view of the relationship between individuals and the food supply. Significantly, Wallace too recalled his reading of Malthus at a crucial point in his process of discovery, but he came from a very different social background from Darwin and may have read different implications into the idea of population pressure.

Functionalism

Darwin's model of evolution was not only gradualist and based on individual actions, it was also conceived as a way of explaining how species become adapted to their environments. If natural selection were the only mechanism of evolution, then every characteristic of every species must either be adaptive or be the remnant of something that had an adaptive function at some point in the past. Modern neo-Darwinists accept this implication, but it is by no means the only foundation for a theory of evolution. Many naturalists believed that species possessed characteristics totally unrelated to the demands of adaptation and hence incapable of being formed by natural selection (or indeed simple Lamarckism). They assumed that purely internal, biological forces shaped the development of individual organisms along predetermined paths that could not respond to environmental influences. The demands of adaptation might dictate the superficial aspects of a species' structure, but its deeper internal form was predetermined, and the deeper structures were responsible for the resemblances between species that allow them to be classified into coherent groups.

This structuralist tradition was already well established in Germany during Darwin's time. The nineteenth century was racked by the great debate over the relative power of form versus function to determine biological structure.[23] The formalists or structuralists believed in internal and hence non-adaptive control, while the functionalists assumed that adjustment to the environment (past and present) explained both the variety of form and the more basic relations between species. The great French anatomist Georges Cuvier adopted the functionalist viewpoint, while his rival Étienne Geoffroy Saint-Hilaire promoted structuralism. The German followers of Blumenbach also favored the formalist approach. Formalists could accept evolution by supposing that forces from within an organism controlled variation; indeed, Geoffroy proposed a theory in which new species were formed by what we would call macromutations. Cuvier believed the similarities by which species are grouped reflect deeper levels of functional efficiency beyond those of adjustment to the local environment, a view that Darwin would replace with the supposition that relationships derive from common ancestry in an earlier form that had itself been adapted to a past environment.

Darwin remained true to functionalism, convinced that the main task of an evolution theory was to explain how species became adapted to their environments. His decision can be seen as a reflection of British naturalists' traditional commitment to the functionalist version of the "argument from design" embodied in William Paley's *Natural Theology* of 1802, a book that fascinated the young Darwin. Paley's argument was summed up in his classic example of the watch and the watchmaker. If you see a watch, you know that an intelligent artificer constructed it to serve a particular purpose. So when you see the complex structures of living things, all adapted to serve the needs of the animals that possess them, you have to ask how this intelligent design (to use the modern term) came about, and the only answer is the Creator. It was this version of what we now call creationism that Darwin set out to challenge by suggesting adaptation as a natural process rather than as a fixed state indicating supernatural design. And because Paley focused on adaptation as a sign of the Creator's benevolence, adaptation became the central focus of Darwin's theory.

Paley's version of the argument from design resonated with the free-enterprise ideology that was a crucial foundation of Darwin's thinking. Paley's thesis is sometimes called the "utilitarian" argument from design because it focuses on the usefulness of each adaptive structure. And utilitarianism was a key concern of the political ideology of the rising commercial and industrial classes: the whole point of any activity was to create wealth and happiness by being *useful*. Here again we can see why only someone operating within the cultural environment of early to mid-nineteenth-century Britain could have developed the theory of natural selection.

The rival structuralist approach has largely been written out of the history of biology in the English-speaking world. Applied to the history of life, structuralism promoted the vision of parallel lines of predetermined evolution, often modeled on the process of embryological development (for which reason I call it the "developmental" alternative). This view was active in France and Germany long before Darwin published, and it survived in the non-Darwinian theories of the later nineteenth century. Yet—following Darwin's own lead—British and American scientists and historians have tended to see the rise of evolutionism as a battle between Darwinism and Paley's version of creationism. Darwin wanted his readers to believe that there was no alternative between a biblical vision of creation and his own theory, thereby effectively leaving out the rival developmental tradition. To some extent this reflected the situation in Britain, where structuralism was only slowly introduced from the Continent during the mid-nineteenth century. But to take a wider view—and to understand what might have happened if Darwin had not written the *Origin*—we need to appreciate that this very non-Darwinian vision of nature had a real presence in science at the time. Even in our own world it flourished into the late nineteenth century and has reemerged in modern evolutionary developmental biology. Without Darwin, it would have played an even more prominent role in the emergence of an evolutionary perspective.[24]

The Animal Breeders

We have now established all the foundational assumptions needed to formulate the theory of natural selection—and we have seen how dependent those assumptions were on particular scientific and cultural attitudes. But to put all these pieces together required a distinct act of conceptual innovation. Darwin had to recognize that a population composed of varying individuals could remain adapted to a changing environment if those individuals whose characteristics best fitted them to the new situation reproduced more prolifically, while those least well-adapted reproduced less. For him—but, crucially, not for Wallace—this insight was inspired by his work with animal breeders, the only area (at the time) in which changes could actually be seen within species on a human timescale.

Darwin became so enamored of the analogy between artificial and natural selection that his terminology—"*natural* selection"—implied that his discovery was a direct recognition that there could be a natural equivalent of the breeders' activities. Historians working on his notebooks from the late 1830s now think the inspiration was less direct, but there can be little doubt that the decision to study the breeders' methods gave Darwin both a fund of information about variation and heredity, and a model that shaped his thinking about how nature might operate.[25] Again we see the contingency of the process of discovery. Darwin was a trained naturalist with experience of biogeography and a commitment to the uniformitarian methodology that encouraged him to look even to an artificial process for information about the way animals could change on a day-to-day basis. He also became a country gentleman who could easily follow up his inclination to interact with the communities of pigeon fanciers, dog breeders, and other groups who knew that selection of individual differences in every generation was the key to success in the production of a new characteristic.

The breeders could produce new characteristics in their tame populations because they could determine which animals reproduced, preferring those with any slight variation in the desired direction. In practice this often meant that they decided which individuals should stay alive to breed—the others were simply destroyed. Darwin's insight was that the designing hand of the breeder was mirrored by nature because those individuals with a variation useful to themselves would be healthier and better fed and would tend both to survive and to breed more successfully than those with less well-adapted variations. The latter would probably simply die off. The useful or adaptive characteristic would thus be enhanced in every generation until it eventually became the norm for the whole population.

This was the basis for natural selection, and throughout his career Darwin appealed to artificial selection as a model to help his readers visualize how the process worked. Wallace, in contrast, was in no position to investigate breeders' methods because he was working in the Far East until 1862, four years after he conceived the idea of natural selection. Moreover, when he learned the details of Darwin's theory from the *Origin*, he was suspicious of the analogy between artificial and natural selection. He even distrusted the term "natural selection" because it seemed to imply that there was an intelligent force at work in nature analogous to the designing hand of the breeder. He was quite right, and Darwin found that he was frequently misunderstood by people who thought that nature was a benevolent agent working for the species' good.[26] This may have helped to minimize public antagonism by concealing the true nature of the theory—in fact, there is no benevolent agent at work, and the whole process is an essentially selfish one based on the reproductive success of those who have some advantage over their neighbors. Wallace's concerns were justified, but when we come to consider his own discovery, we shall have to ask whether his lack of interest in artificial selection may have led him to conceive the natural process along lines significantly different from those followed by Darwin.

The Struggle for Existence

On one component of the theory Darwin and Wallace did agree. In principle, natural selection should work under any circumstances in which the fittest—that is, better-adapted—individuals breed more successfully than other members of the population. Even in a world of unlimited resources, where the unfit could still manage to survive, the fit would still tend to breed more often and thus enhance their characteristics in the next generation. But for Darwin in 1838 (and Wallace twenty years later), this does not seem to have been enough. Neither man conceived selection as an effective mechanism until he saw it as something driven by a more powerful force, and that force was the shortage of resources arising from population pressure—the struggle for existence. This insight came in both cases from reading Thomas Malthus's *Essay on the Principle of Population* (for Wallace, it was the memory of reading the essay much earlier). Malthus set out to undermine ideas of social progress based on the redistribution of wealth by arguing that poverty is natural and inevitable. Like all other animals, humans tend to reproduce more rapidly than would be needed to maintain a stable population. The result is pressure by a potentially expanding population on limited environmental resources. More individuals are born than can be fed, so many must die, and the result is a competition for scarce resources, which Malthus called the "struggle for existence." Darwin and Wallace both saw this as the driving force of selection.[27] The unfit would not only be less capable of breeding, they would lose out in the struggle and be eliminated from the population.

Malthus's input forms the key debating point in the dispute over whether or not the selection theory is a projection of the free-enterprise value system onto nature. Malthus, like Paley, was a product of the individualist, utilitarian ideology. Both saw the pressure of population as a divine ordinance imposed to reward the virtues of thrift and industry. The only way of overcoming the natural consequences was to teach people the wisdom of refraining from producing children they could not support. The last thing the state should do is attempt to alleviate the poverty of those who do breed without making proper provision, because their children too would go on to breed, and soon there would not be enough food to go around, whatever the good intentions of those in charge. Curiously, Malthus only used the term "struggle for existence" when he discussed warring tribes in central Asia, not when he explored the consequences of his theorem for the individualistic society of industrializing Britain. But Darwin saw immediately that the message applied as clearly to individuals as to tribes. Just as in the breeder's flock of pigeons, it was the favored individuals who got to survive and breed. Darwin unpacked consequences that Malthus himself did not see in his principle, for all that he discouraged any attempt to interfere with nature's harsh laws by public charity.

Malthus's book had been published in 1798 and by Darwin's time had expanded into a hugely detailed tract. There had been an impassioned debate over its implications; the reformers and revolutionaries hated it, while those who shared its utilitarian values applauded this addition to the "dismal science" of political economy. Darwin came from the class that was inclined to take the harsher viewpoint, although for a naturalist to look to political economy for inspiration was highly unusual. Contrary to his later claim that he read Malthus "for amusement," he came to it as part of a program of reading designed to help him think through the consequences of applying his ideas to the human race. Here Darwin's intellectual radicalism combined with a political conservatism to put him in a unique position. Only he could appreciate how Malthus's principle might be linked to the study of biogeography and his work with the breeders. No one else was in a position to pull all of these scientific and cultural factors together in 1838 to form the basis for a comprehensive theory of evolution. He then undertook twenty years of work on the topic before he wrote the *Origin of Species*. When we factor this study period into the equation, it is hard to see how anyone else could have been prepared to write such a book at the time.

THE PRECURSORS

We now have to confront the efforts made by some historians to challenge Darwin's achievement by claiming that there were precursors or co-discoverers who independently conceived the theory. If this claim proves true, my case for seeing Darwin as a unique thinker is weakened. This is a sensitive topic because the various alternative discoverers each have a retinue of modern supporters determined to advance their hero's claim by arguing that Darwin's status has been exaggerated. There are charges of a conspiracy by Darwin and his supporters to marginalize the rivals, and even a claim that Darwin actually plagiarized part of the theory from Wallace.

There is more going on here than a simple priority dispute. Some of the efforts to claim that Darwin was not the real discoverer of natural selection are based on an oversimplified application of modern standards for assessing priority of discovery and publication. In a world where every scientist seeks to make his or her reputation by announcing a new discovery as quickly as possible, usually through a periodical article or online, it is hard to imagine how different things were in the mid-nineteenth century. Major new ideas were often still promoted in books, and in the case of really revolutionary proposals, it was necessary to provide this level of support to ensure that the theory would not be dismissed out of hand. An outline of the basic idea of natural selection can be provided fairly concisely, as the Darwin and Wallace papers of 1858 demonstrate. But to make a case for the theory of divergent evolution by natural selection in a way that would force everyone to take it seriously, a much more substantial discussion was needed. In these circumstances, application of a rigid standard of priority in publication is hardy appropriate. Patrick Matthew may well have stated the idea of natural selection as early as 1831, but he did nothing to explore its implications or to persuade his readers that it had the potential to revolutionize biology. His contribution is worth noting, but to suggest that it provides the basis for dismissing Darwin as the true founder of the theory is to misunderstand the whole process of how a scientific revolution happens.

The real issue here is that Darwin's detractors often seek to undermine his claim by blackening his reputation. It is perhaps fair to say that Darwin has become a figurehead of the evolutionary movement to such an extent that he has obscured the contributions of many other naturalists (including many who have no claim to discovering natural selection). But the enthusiasm of those who seek to erode Darwin's reputation makes it hard to engage in rational dialogue on the issue. Historians looking for a more nuanced view have to battle with conspiracy theorists and writers who repeat the same stock arguments with little regard to the complexities of the case. The way in which Darwin and his supporters handled the publication of Wallace's paper in 1858 is routinely cited as evidence of a conspiracy to sideline Wallace, without any investigation into how the latter might have gotten his paper published, considering that he was in the Far East at the time. The charge that the academic community blocks all efforts to publicize the critics' case is repeated over and over again. The claim that Wallace is a forgotten genius has been made in a whole series of books, each of which has to ignore all predecessors to make its point.

Perhaps what we are seeing here is a natural sympathy for the underdog, coupled with a sense that the glorification of Darwin has gotten out of hand. One can sympathize initially with these motivations, but when they are applied in a manner that seeks to undermine the immense originality of the case that Darwin presented in the *Origin of Species*, sympathy begins to evaporate. More to the point, these various claims for pre- and co-discovery lend support to the impression that the idea of natural selection was somehow in the air at the time. All of the components were available, and other people were putting them together, so why should we give Darwin all the credit for something that was pretty obvious anyway? It should already be clear why I do not accept this assessment of the situation. The individual components may all have been in the public domain. But to see how they could all be fitted together and to explore the wider implications of the resulting conceptual package required someone able to synthesize ideas and information from a variety of disparate sources and to think outside the box about their implications. If we run through the claims for the rival discoverers, we can see how they fail in one or more ways to meet these criteria.

The main contenders to be considered as having predated Darwin in the discovery of natural selection are William Charles Wells, Patrick Matthew, and Edward Blyth. Wallace belongs in a separate category, since he is presented as a co-discoverer of the theory who was published alongside Darwin in 1858. The rival claimants certainly show that the idea of the "struggle for existence" had been recognized in the early nineteenth century. But many recognized the harshness of nature without seeing that it might lead to change—after all, Malthus himself had used his principle to argue against the idea of progress. Some, like the philosopher Herbert Spencer, appreciated that struggle might have a positive effect, but saw this operating though a stimulus to self-improvement (essentially a form of Lamarckism). Only a handful of naturalists recognized that the struggle between individuals could lead to a selective effect equivalent to that of animal breeders. And to make full use of this idea, they would already have had to abandon the fixity of species and recognize that individual variation was potentially unlimited, a conceptual revolution that few were willing to make before the 1850s.

Wells and Blyth

Two of Darwin's precursors fall at the first hurdle, since they did not see the process of natural selection as something capable of producing new species. William Charles Wells's "An Account of a Female of the White Race of Mankind, Part of Whose Skin Resembles That of a Negro" was published in 1818 and includes comments on the formation of the human races.[28] Wells notes briefly that a process of selection similar to the one at work in animal breeding may have adapted human populations to the different environments into which they migrated. He does not seem to have followed the suggestion any further, and he made no claim that the process of selection would produce changes sufficient to generate new species. Wells thought the human races were varieties of one species, and his few references to animals mention only artificial breeds.

Loren Eiseley and others championed Edward Blyth mainly on the basis of his 1835 paper, "An Attempt to Classify Varieties of Animals."[29] Here

Blyth's main concern was to clarify the various senses in which the term "variety" was used at the time. In discussing "true varieties," that is, well-established races with distinct characteristics occupying particular territories, Blyth notes the analogy between artificially bred varieties and the human races. Like Wells, he offers the possibility that these variant forms are produced by adaptation to the environment, and he mentions a kind of natural selection caused by the tendency of the strongest individuals to displace the weaker. In the case of wild cattle, the strongest bull drives out his rivals "so that all the young which are produced must have had their origin from one which possessed the maximum of power and physical strength; and which, consequently, in the struggle for existence, was best able to maintain his ground, and defend himself from any enemy." More generally it is always the best-organized individual who will "transmit its superior qualities to greater number of offspring." The process might thus produce new varieties adapted to local conditions, but there is no suggestion that the process could be extended to produce new species. Blyth was in fact convinced of the stability of species, and his version of the theory was conceived within a framework of natural theology. Human breeders make use of "The same law . . . which was intended by Providence to keep up the typical qualities of the species."[30]

Wells and Blyth show that there were at least a few naturalists willing to see an analogy between natural and artificial selection, and both recognized that those best fitted to an environment would tend to breed more successfully. But neither was in a position to see selection as a mechanism of change powerful enough to threaten the plausibility of the traditional view that species were created with fixed characteristics. It is conceivable that Darwin read Blyth and thus might have picked up on his brief exposition, although he makes no mention of him at the key points in his notebooks. But even if Blyth's suggestion had lingered at the back of his mind until it triggered his own inspiration, this was only possible because he could fit the idea into a far more radical theory of evolution.

Matthew

Patrick Matthew's claim is based on firmer foundations, because he did at least appreciate that natural selection was a true mechanism of what we call evolution. His account comes in an appendix to his 1831 book on *Naval Timber and Arboriculture*. Precisely because he was writing about the possibility of improving plant species, especially trees, there is an immediate link with the breeders' use of artificial selection. Matthew recognized that there would be a struggle for existence in nature (although he does not mention Malthus) and that it would be the best-adapted individuals who survive and breed, transmitting their characteristics to the next generation. "As the field of existence is limited and pre-occupied, it is only the hardier, more robust, better suited to circumstances individuals, who are able to struggle forward to maturity, those inhabiting only the situations to which they have superior adaptation and greater power of occupancy, than any other kind; the weaker, less circumstance-suited, being prematurely destroyed."[31] He appreciated that the result would be not only the adaptation of species to changes in their environment, but also the diversification of living forms in the course of geological time.

A significant difference between Matthew's vision and Darwin's is that Matthew linked his to a catastrophist theory of earth history, imagining sudden bursts of evolution branching out from the survivors of a series of mass extinctions. This is perhaps not surprising: in 1831 Charles Lyell had only just published the second volume of his *Principles of Geology*, outlining the rival uniformitarian position that Darwin would adopt. And there is no doubt that Matthew did envision a limited amount of natural selection continuing to adapt species to less abrupt changes in times of restricted geological activity, such as the present.

With his interest in horticultural breeding, Matthew even suggested that experiments might be done to throw more light on the production of new variations. But there is nothing to indicate that he followed up this line of investigation, nor did he publish anything else to show how his idea might be developed into a complete theory capable of overcoming the many potential objections to it and unifying the diverse areas of natural history. His remarks were confined to a few paragraphs in the appendix to his book, which were ignored until, following the publication of the *Origin of Species*, Matthew himself began a campaign to gain recognition for his earlier suggestion of the selection hypothesis.

Darwin conceded that he had been preempted in this case, but he noted Matthew's failure to make any comprehensive use of the idea. Matthew's modern supporters, especially W. J. Dempster, stress that his discovery predates Darwin's earliest investigations and insist that because he also published first he deserves the real credit for the theory.[32] The incident certainly shows that it was possible for others to conceive of natural selection as an evolutionary mechanism, but Matthew's apparent lack of interest in the idea only highlights the fact that mere discovery and publication is not enough to initiate a scientific revolution. Having a basic idea, even publishing it, has no effect if the publication is obscure and no further effort is made to exploit and promote it. Darwin and Wallace's joint publication of 1858 was similarly ignored, but Darwin was armed with his twenty years of experience and was able to produce a more substantial text the following year that forced everyone to sit up and take notice. We talk about a "Darwinian revolution" because it was Darwin, not Matthew, who initiated the transformation in our way of thinking about the world.

WALLACE

Alfred Russel Wallace is a very different case. He certainly doesn't provide an instance of simultaneous discovery, because he conceived the idea of natural selection twenty years after Darwin. But far more than Matthew he represents someone who was working toward a general theory of adaptive evolution and was thus in a position to appreciate the potential significance of the idea. The peculiar circumstances surrounding his discovery and its eventual publication alongside an abstract of Darwin's own work have generated endless controversies. Wallace's supporters generally concede that he was following in Darwin's footsteps, but they argue that because the latter was not about to publish under his own steam, Wallace was shortchanged in the arrangement made to publish the joint papers.[33] He appears as the second in line, whereas he ought to have been presented as the primary initiator of the publication. Other critics go much further. Arnold Brackman even suggested that Darwin plagiarized the idea of divergence from Wallace, covering up the true date of arrival of the latter's 1858 paper.[34] Few historians took this claim seriously, and John Van Wyhe and Kees Rookaaker have since decisively refuted it.[35] But the belief that Wallace was badly used by Darwin's supporters has attracted a band of sup-

Figure 3 · Alfred Russel Wallace as a young man.

porters determined to rescue him from his alleged obscurity in Darwin's shadow. Michael Ruse has somewhat unkindly referred to them as 'Wallace groupies.'[36]

Significantly, Wallace himself did not challenge the way his paper was handled by Darwin's supporters and always accepted that he had played a less significant role in the theory's development. He was quite happy to accept the term "Darwinism" and even used it as the title of one of his later books. In a recently discovered letter written to Charles Kingsley in 1869, he compared himself to a guerrilla chief who might win a skirmish, while Darwin was the great general who could lead an army into battle.[37] Wallace did not see himself as someone who could have initiated a scientific revolution—although this does not mean that he would have played no role in a world without Darwin.

Wallace's Discovery

Wallace came from a relatively poor background and became a traveling naturalist, exploring first in South America and then in the Malay Archipelago (modern Indonesia), supporting himself by selling duplicate specimens of rare species to collectors back home. His studies of biogeography led him toward the idea of transmutation, a move encouraged by admiration for Lyell's approach to geology and his reading of Chambers's *Vestiges* (of which he had a much higher opinion than Darwin). In 1855 he published a paper noting that new species always appear in an area occupied by a previously existing, closely related species. He used the model of a branching tree as a way of explaining the relationships between species, exactly paralleling Darwin's first insight into evolutionary relationships. Lyell was impressed, but Darwin was not, because Wallace did not explicitly state that the most obvious implication of his argument was that the new species were derived from old ones by transmutation. Wallace's key inspiration came in 1858 while he was suffering a bout of fever. He remembered an earlier reading of Malthus, recognized the significance of the struggle for existence, and saw that as a consequence the best-adapted varieties would survive and breed while the least well-adapted would eventually die out. He wrote this idea up in a short paper and sent it to Darwin, whom he knew to be interested in the topic, to see if he thought it was worth publishing.

Darwin read the paper, saw the similarities to his own theory, and panicked, fearing that he would lose his twenty-year priority. He called in Lyell and Hooker, who advised that Wallace's paper should be published alongside a short account by Darwin himself, including material that could be independently confirmed as having been written before Wallace's discovery. The joint papers of 1858 thus included an extract from a letter that Darwin had written to the American botanist Asa Gray in 1857. Wallace continued to work on the theory after his return to England in 1862 and published important studies of speciation and geographical distribution, while engaging in a protracted debate by letter with Darwin over the details of how the theory should be developed.

Two Versions of Natural Selection?

Wallace's contribution raises a series of issues relevant to the question of Darwin's influence. Was this really a case of independent discovery of an identical theory? Wallace's modern supporters certainly feel that he had conceived the full theory of natural selection, and some claim that his initial formulation was superior to Darwin's. The possibility that Wallace's 1855 paper helped Darwin to understand how natural selection drives species toward divergent specializations does at least merit some attention. This point does not, however, entail that Wallace's 1858 paper contained a clear description of the theory of natural selection acting on individual variants within a population. The Wallace enthusiasts take this interpretation of the paper for granted, as do all advocates of the "independent discovery" thesis. But there is a longstanding tradition that throws doubts on this reading, noting some significant differences between the ways the two men seem to have thought selection would operate. It is possible that, left to himself, Wallace would have promoted a significantly different evolutionary theory from the one Darwin published in the *Origin of Species*—in effect Wallace "Darwinized" his own thinking after reading Darwin's book. There are also more practical concerns. In a world without Darwin, to whom would Wallace have sent his paper in the hope of publishing? Indeed, would he have had any hope of publishing at all? Even if his paper were published, would it have had any effect (given that the joint papers attracted very little attention in our world)? To have an impact equivalent to that of the *Origin*, Wallace would have had to write a substantial book of his own, and this would have taken him many years.

The most controversial of these issues is the question of whether Wallace's 1858 insight really parallels the key points already established by Darwin. Wallace's supporters adopt Darwin's own original perception, which assumes that the 1858 paper contains the whole theory. But there is an alternative reading of the paper first suggested by the American palaeontologist Henry Fairfield Osborn in 1894, noted by the biologist Edward Bagnall Poulton in 1896, and maintained currently by a number of historians (myself included). In the volume published to celebrate the 1959 centenary of the *Origin*, A. J. Nicholson suggested that while Darwin always visualized selection in terms of competing individuals seeking to survive in a given environment, Wallace seemed to think that the environment sets a fixed standard against which organisms are measured, some passing the test and others failing. This reading parallels Osborn's interpretation, which I revived (unknowingly) in 1977, based on the observation that in his 1858 text, Wallace's use of the term "variety" is surprisingly ambiguous.[38]

The title of Wallace's paper was "On the Tendency of Varieties to Depart Indefinitely from the Original Type," and Osborn's point focuses on the meaning of the term "variety." Wallace argues that the survival or elimination of "varieties" is determined by their adaptive character, and the traditional assumption is that he was referring to the individual variants within the population that are the basis of Darwin's proposed mechanism. But the term "variety" was normally used to designate a coherent local population with a distinct character adapted to its own microenvironment, that is, a group of organisms all sharing the same peculiarity. Much of Wallace's text can be read as a description of a kind of group selection. Species somehow divide themselves into a series of local populations (varieties), and eventually one of those populations turns out to be better adapted to the whole territory occupied by the species, so it expands and takes over the territory, wiping out the others varieties in the process.

Darwin accepted that closely related varieties and species were in competition, but for him this was not the most basic level at which selection acts. The core Darwinian mechanism is a process of selection *acting on individuals within a single population*, and this is responsible for dividing the species up into distinct varieties on the way toward becoming new species. Did Wallace recognize this level of selection in 1858? His reading of Malthus certainly pointed him toward competition at the individual level, but much of his text is ambiguous on the point, and some passages openly portray selection in terms of one population replacing another. Thus when conditions change: "it is evident that, of all the individuals composing the species, those forming the least numerous and most feebly organized variety would suffer first, and, were the pressure severe, must soon become extinct. The same causes continuing in action, the parent species would next suffer, would gradually diminish in numbers, and with a recurrence of similar unfavourable conditions might also become extinct. The superior variety would then alone remain, and on a return to favourable circumstances would rapidly increase in numbers and occupy the place of the extinct species and variety."[39] Phrases such as "become extinct" and "occupy the place of" suggest that Wallace is thinking of competing groups, not individuals. He may have recognized the existence of individual natural selection, but his attention was at this stage focused far more strongly on what was, to Darwin, a secondary level of selection operating between distinct populations. Several modern historians of evolutionism, including Janet Browne and Michael Ruse, accept this interpretation of Wallace as (initially, at least) a group-selectionist.[40]

In a letter to Poulton, Wallace subsequently insisted that he had used the term "variety" only because individual variation was hardly recognized at the time, and that he had meant it to include the latter phenomenon.[41] It cannot be denied that Wallace soon came to appreciate the importance of individual selection—but this was *after* he read Darwin. Significantly, when he reprinted the 1858 paper in his *Contributions to the Theory of Natural Selection*, he added section titles to the original text, at least one of which encourages the reader to think of varieties as individual variants: "Useful Variations will tend to Increase; useless or harmful Variations to Diminish."[42] These subtitles have been retained in almost all subsequent reprintings of the paper, helping to perpetuate the impression that the original text was primarily about individual selection.

Once he had read the *Origin* and argued relentlessly with Darwin over the details of how the mechanism operated, Wallace soon came to grips with the concept of individual selection. He exploited the theory in his own research and in some respects came to a much clearer conception of individual variation than Darwin himself. In his book *Darwinism*, Wallace depicts the range of variation in a wild population using something very close to the bell curve of modern statistical analysis.[43] But the crucial question for counterfactualism is, given the ambiguities in his 1858 paper, how would Wallace have developed the idea if there had been no Darwin for him to read? It can plausibly be argued that he would have continued to focus on group selection, using this to explain the geographical dispersal and multiplication of species, which was, in fact, one of his main interests and—even in our world—his main contribution to science.

The idea of selection operating at the individual level would thus have remained in the background and would not have been thrust onto the public's attention. The harsher implications of the full Darwinian theory would have been masked to some extent, although the effects of the struggle for existence in eliminating the less fit varieties and species would still have been apparent. This last point is important, because in the late nineteenth century, the effects of this level of selection were widely accepted even by scientists and other thinkers who did not appreciate the significance of selection acting at the individual level. In our own world many anti-Darwinian evolutionists accepted group selection as a negative process eliminating the less efficient products of evolution, while insisting that a more purposeful force must actually produce the new forms.

Two differences between Darwin and Wallace support the claim that the latter would not have developed a full theory of individualistic selection. One is that Wallace made no appeal to an analogy between natural and artificial selection in 1858 and remained consistently suspicious of its validity throughout his long interaction with Darwin. His work in the 1860s focused on natural selection in the wild and made no mention of animal breeding. As late as 1889, his survey *Darwinism* contains only a short, eighteen-page chapter on artificial selection, appealing to the breeders mostly for evidence of a significant range of variation within populations. Yet for Darwin, the analogy with the breeders' work was always a key explanatory strategy, and, crucially, it focuses the readers' attention firmly on selection operating between individuals. Wallace's persistent reluctance to acknowledge a parallel with the breeders' method is both consistent with the claim that he was not thinking in individualistic terms in 1858 and an indication that any theory he developed on his own would have lacked a central theme of the *Origin of Species*. Wallace had valid arguments on his side—he disliked the term "natural selection" because it encouraged readers to see the process as governed by an intelligent selecting agent, an impression reinforced by drawing an analogy with the work of the breeders. Selection was a bad metaphor for a process that was in fact purely mechanical. But it was the power of this metaphor that impressed Darwin's readers and forced them to focus on the process of individual selection, even if they did read implications into it that Darwin did not intend. If Wallace alone had launched the theory, the lack of this powerful image, coupled with Wallace's own tendency to focus more on group selection, would have marginalized key aspects of what we call Darwinism.

The second difference between the two men lies in their wider beliefs.[44] Wallace came from a poorer background than Darwin and was politically much more radical. He would not have been attracted to a version of the theory based on an analogy to the competitive ideology of free-enterprise capitalism. Wallace was also deeply religious, a point that came to the fore when he began to argue that the higher qualities of the human mind could not have been evolved by natural selection and needed some supernatural input. For as much as he proclaimed natural selection to be the only mechanism of animal evolution (rejecting Darwin's own inclusion of an element of Lamarckism), Wallace's last book on evolution, *The World of Life* (1911), argues that the history of life on earth represents the unfolding of a divine plan. Throughout his career, he strove to minimize the harsher implications of the theory. He insisted that animals do not feel pain as we do, so there was no need to see nature as harsh and cruel because the unfit had to be eliminated. Tennyson's famous line about "nature, red in tooth and claw" was, he insisted, an inappropriate projection of human experience onto the animal kingdom.[45] He made few if any references to parasites or to the indifference of predators to the suffering of their prey, all factors that represented to Darwin the cruelty and selfishness of nature. A theory developed and presented by Wallace alone would have been much less likely to shock his readers and to challenge their deeper beliefs.

Wallace would also have faced difficulties at a practical level. When he wrote his 1858 paper, he was in a remote location in the Far East. He sent his manuscript to Darwin because he had heard that the latter was working on the species question. But if there had been no Darwin to send it to, what would Wallace have done with his paper? Almost certainly he would have sent it to Lyell, but there is no guarantee that it would have received an enthusiastic reception. Lyell had been impressed by Wallace's 1855 paper and suspected that it might imply transmutation, but he was deeply troubled by the idea. Even after being primed by Darwin, he only reluctantly adopted evolutionism in the later 1860s. In a world without Darwin, Wallace's 1858 paper would have come as quite a shock, and there is no way of telling how Lyell would have reacted. He might have ignored the paper as too radical or sent back a list of queries and objections. Even if Wallace had been able to publish, we have to bear in mind that in our universe the joint Darwin-Wallace papers attracted very little attention. How much less would be the interest in a single short paper written by an unknown naturalist collecting in the jungles of the Malay Archipelago? It was the *Origin of Species*, not the 1858 papers, that precipitated the great debate, so in a world without Darwin, Wallace himself would have had to produce a text of equivalent weight if he were to be taken seriously. Since he did not return to England until 1862, it is hard to imagine him beginning this task before then, and it's unlikely that he would have produced a major manuscript in anything less than five years. Wallace's serious initiative would have been launched sometime in the late 1860s, perhaps even in the 1870s. By this time the whole situation would have changed.

Darwin was the only person in a position to launch a major initiative based on the idea of natural selection in 1858–59. The "in the air" thesis fails because of the very small number of alternative discoverers and the major differences between the ways they conceived the theory. But natural selection was not the only game in town, and other developments were beginning to promote a range of evolutionary alternatives. Darwin himself sensed that attitudes in the mid-1850s were changing enough to make the general idea of the natural production of species more acceptable, even in Britain. By this time the general idea of evolution really was "in the air," and if Darwin had not published, there were others who would eventually have tried their hands. But they would have been promoting nonselectionist mechanisms such as Lamarckism, and these would have become the foundations of the first evolutionary paradigm. If Wallace had had to wait until around 1870 to publish a substantial account of his theory, he would have struggled to gain a hearing within a debate whose parameters were already being defined by very different models of evolution. To visualize the foundations of this hypothetical debate, we only have to look at what happened in our world while Darwin was working in secret until 1858.

3

SUPERNATURALISM RUNS OUT OF STEAM

Darwin wrote a substantial account of his theory in 1844 but made no effort to publish it. Historians have assumed that he was afraid of the public reaction, especially as his wife, Emma, had concerns about the implications of his idea. 1844 was the year in which Chambers's *Vestiges* appeared, and the conservative elite of the scientific community recoiled in horror from its suggestion that humanity was the product of a natural process of development. Historian John Van Wyhe has pointed out that Darwin never explicitly said that he delayed publication out of fear, noting that for the next ten years he was busy with other projects. In the aftermath of the *Beagle* voyage, he was still working up his geological observations, and soon he began his enormously time-consuming study of the barnacles. He was also gathering evidence that might help him to make a more convincing case than was outlined in the 1844 essay. These practical concerns certainly demanded his attention. But most Darwin scholars suspect that he felt at least a subconscious fear of the public reaction, which prompted him to find excuses for putting off the publication of his theory.[1]

When Darwin conceived his theory in the late 1830s, most British naturalists thought that evolutionism was a danger to religion, morality, and the social order. It was also considered bad science. That many still felt this way in the mid-1840s is evident from the reaction to *Vestiges*. But the attitude of many French and German naturalists was less negative, and there is evidence that in Britain, too, support for the idea of miraculous creation diminished in the 1850s. Darwin began to reconsider his policy of secrecy, prompted in part by Hooker and Lyell, the two friends to whom he had unburdened himself (Hooker was told about the theory in 1844, Lyell in 1856). Was this purely because he had now finished the barnacles and gathered enough evidence to make a secure case for the theory? Or do the actions of the three scientists suggest that they had detected a change in the public attitude toward evolution, a change that would allow a new initiative to be launched without attracting the hostility endured by *Vestiges*?

A new vision of nature was emerging that would indeed provide a more positive climate for discussion of evolution. Even though hardly anyone knew about Darwin's theory of natural selection, the prospect of a naturalistic explanation of the origin of species was now in the air. Darwin's decision to begin writing his "big book" (interrupted in 1858 by the arrival of Wallace's paper) was a product both of his growing confidence in the case he could present and of his recognition that he would now face a less hostile public and scientific community.

This new attitude can hardly have been a product of Darwin's own activity. Although he was in touch with a wide community of naturalists, he had been very slow to spread the word about his new idea. From 1844 to 1856 Hooker had been his only confidant; in 1856 he told Lyell, and, in the following year, he told Asa Gray. Huxley knew only that Darwin no longer accepted the fixity of species—he did not know about natural selection until it was published. The fact that Darwin kept so quiet about his idea thus allows us to use what we know about changing opinions in our own world to construct with some confidence a prediction about the state of affairs in a world without Darwin. There would have been a general swing of opinion away from the traditional idea of the miraculous creation of species, as well as a growing willingness to consider some form of naturalistic alternative. Supernaturalism was on the wane, and the search was on for a natural explanation—although the theory of transmutation was only one possibility under consideration.

In our universe, the *Origin of Species* provided the stimulus for a wholesale transition to evolutionism (most naturalists ignored the 1858 Darwin-Wallace papers). Evolution in its most radically naturalistic form became a central feature of debate. But without the *Origin*, would there have been the incentive to reopen a controversy that had lain dormant since the furor over *Vestiges* fifteen years earlier? We have been conditioned to believe that Darwin's book came like a bolt from the blue that completely transformed the situation. Unless we invoke an equivalent stimulus from Wallace—which I have shown to be most unlikely—perhaps the whole idea of evolution would have remained on the back burner for years and there would be no equivalent of the Darwinian revolution.

T. H. Huxley is in part responsible for the image of Darwin's book as a transforming influence. As Adrian Desmond has noted, Huxley's "Coming of Age of the *Origin of Species*" of 1880 gave the impression that until 1859 virtually everyone thought of the history of life as a sequence of catastrophic mass extinctions followed by bouts of supernatural creation. Another of Darwin's leading supporters, Ernst Haeckel, later argued that most German scientists were hostile to any form of evolutionism in the 1860s. But these assessments cannot be taken at face value. Both men had much to gain from highlighting their own contributions to the revolution that Darwin had precipitated. Huxley had in fact been reluctant to consider evolution before 1859 and was suspicious both of the adaptationist project to which Darwin was contributing and of the theory of common descent. As he later reminisced, he had been inclined to say to both creationists and evolutionists, "a plague on both your houses." If this attitude was general (as Huxley implied), then without the stimulus provided by Darwin's new hypothesis about the mechanism of change, the whole evolutionary project might have remained in abeyance.[2]

It's possible, though, that Huxley and Haeckel were exaggerating the extent of scientists' reluctance to consider revolutionary ideas about evolution. The claim that almost everyone was still an extreme catastrophist is certainly false, because a number of eminent scientist and intellectuals had begun to drop hints that an alternative to the idea of divine creation was needed. But this still leaves open the possibility that without Darwin's suggestion of a new hypothesis about the cause of evolution, scientists would have remained reluctant to reopen the case. Pressure for change was building, but the move toward evolutionism was blocked by the inability of the scientific community to develop a plausible theory of how the process worked. Vague ideas about "laws of development" were not enough to turn evolutionism from a pretty speculation into a scientific theory, and without the support of scientists, public debate on the topic would remain muted. Huxley's "a plague on both your houses" comment suggests that there was a kind of deadlock on the topic that could only have been broken by Darwin's theory.[3] But perhaps there were other forces at work that would have moved things forward without Darwin's intervention. Other naturalists, mostly in complete ignorance of what Darwin was doing, were suggesting a range of alternatives, and all pointed toward a rethinking of the traditional view that species were fixed entities whose origin could not be explained by natural law. Wider changes in Western culture were also questioning ideas based on supernatural interference with the world.

Even without Darwin's input, by the 1860s naturalists and intellectuals would have moved far enough away from the traditional worldview of Genesis to reconsider the case for evolution. The blockage in scientists' thinking about species may have been less rigid than Huxley implied, allowing it to be dismantled in stages by less dramatic interventions. Evolutionism would eventually have flourished—but it would have been an evolutionism based on non-Darwinian ideas, not on natural selection.

This position requires a reconsideration of exactly what the *Origin* contributed in our own world. If we can analyze that contribution's component parts, we shall be in a better position to understand what might have happened without Darwin. Historians usually point to the question of the mechanism of evolution as the really key innovation. Lamarck's theory of the inheritance of acquired characteristics had been available for some time, but Huxley and many others regarded it as a discredited idea that was not worth reexamining. Huxley was also a leading opponent of Chambers's vague notion of a law of development, largely because it conceded too much to the argument from design by implying that the direction of change was divinely preordained. Natural selection was crucial for Huxley not because he thought it was an adequate explanation but because it showed that new hypotheses could fill in the gaps that had traditionally left room for the supernatural. New topics such as artificial selection suddenly became relevant, opening the way to a much broader synthesis. Yet Darwin's synthesis did not command assent at the time, and nonselectionist theories soon began to flourish. So the new theory in the *Origin* may not have been as decisive as Huxley implied. There were, for instance, already signs of renewed interest in the Lamarckian mechanism, not least from Huxley's friend Herbert Spencer.

But Darwin's impact went beyond the question of mechanism. He provided a huge battery of facts to support the general case for evolution by common descent. He summarized various lines of evidence in order to throw doubt on the belief that species were clearly defined and permanent entities. Biogeography helped in this respect and also provided the foundation for the model of divergence from a common ancestor. Here we can see more clearly how, in a world without Darwin, the case for evolution would eventually have been reopened, because we know that other naturalists including Hooker, Gray, and, of course, Wallace were becoming increasingly conscious of where the evidence was pointing. This general evidence for adaptive evolution and common descent might have triggered a reconsideration of the Lamarckian mechanism, which would have been invoked to explain the process—as indeed it was by a battery of anti-Darwinian evolutionists in our own world.

We should also bear in mind that not everyone was looking for a theory based on adaptation and common descent. The rival functionalist tradition encouraged alternatives that saw the emergence of new forms as the result of preordained laws. Some naturalists within this tradition still thought new forms of life might emerge directly from unformed matter. Formalism also encouraged a search for inbuilt variation trends that could drive evolution in predetermined directions, a model active in our own world during the eclipse of Darwinism in the late nineteenth century. Conventionally minded naturalists who wanted to see evolution as the unfolding of a divine plan could easily adopt this approach, which was promoted, at least on the surface, by Chambers's *Vestiges*. A significant number of naturalists were already thinking along these lines in 1859, and without Darwin this anti-adaptationist approach to evolution would have flourished and might even have counted Huxley as one of its leading champions. The theory of orthogenetic evolution—naturally imposed variation trends—might have replaced the divine plan of creation. In a world without Darwin, the combination of Lamarckism and orthogenesis, promoted in our world as an alternative to the selection theory, would have become the original foundation for an evolutionary movement in the late 1860s or early 1870s.

At this point, we need to get an overview of just how far the community of naturalists had advanced toward evolutionism by 1859. This requires an appreciation of the scientific discoveries and innovations that paved the way for a systematic exploration of the origin of species. But this exploration involved far more than a scientific revolution, and new ideas about evolution could not have been conceived or promoted unless there had been parallel transformations in the way people thought, especially about the relationship between God, nature, and humanity. Evolutionism threatened traditional beliefs and values, but cultural forces hostile to the authority of the churches were already undermining those traditions. Historical studies of the Bible, coupled with a growing moral revulsion against the belief that nonbelievers faced an eternity of torment in Hell, encouraged a less literal interpretation of the Genesis story of creation and original sin.

Some of the new attitudes we routinely associate with Darwinism, especially the growing enthusiasm for a vision of nature as a scene of struggle and conflict, were being promoted long before Darwin published. In our world, those attitudes were certainly reinforced by what Darwin wrote, but they did not originate with him, and we cannot simply assume that the ideology of social Darwinism would not have emerged in a world that remained ignorant of the selection theory. The Darwinian revolution was not just an event in science—it was a cultural revolution that transformed people's worldview in many ways. But precisely because the transformation was so broadly based, we must not to fall into the trap of thinking that every aspect of the Darwinian viewpoint could only have originated from the *Origin of Species*. The specter of Malthus and the struggle for existence was likely to emerge whether or not Darwin incorporated those elements into a theory of evolution by natural selection.

ALTERNATIVE COUNTERFACTUALS

If other scientists had shared Huxley's skepticism about the possibility of scientific evolutionism, an evolutionary movement might never have emerged without Darwin. But other initiatives in science and elsewhere were paving the way for some form of evolutionary thinking. The main purpose of this book is to explore what these other approaches might have produced if Darwin hadn't been there to publish the *Origin of Species*. But what if he survived long enough to produce the substantial outline of his theory that he wrote in 1844? After all, instructions in his will specified that his wife, Emma, should get it published if he died unexpectedly—by this time he was already experiencing the ill health that would plague him for the rest of his life. Imagine a situation in which he died before writing the *Origin of Species* but the 1844 essay was published instead, perhaps with the aid of Lyell and Hooker. Would we then have had a Darwinian revolution more or less as we know it, but beginning a few years earlier?

The possibility of this "advanced" Darwinian revolution depends on the exact timing of events, and it reminds us that Darwin had an impact through his informal contact with other naturalists as well as through his publications. In the late 1840s Hooker knew about Darwin's theory but was not yet convinced of its validity, and Huxley's career as a professional scientist had not yet begun. The very negative response of some eminent scientists to *Vestiges* would still have been in everyone's mind. Publication at this point may not have been effective, although it is hard to predict what might have happened. But if we imagine Darwin dying in the mid-1850s, the possibility of a campaign to promote his theory posthumously looks much more promising. Hooker was by now a convert, and Lyell was starting to waver, while others such as Asa Gray and Huxley, who was beginning to carve out his career as one of the new breed of professional scientists, were slowly being brought into the Darwinian circle. Although they did not know about natural selection, many naturalists had become aware that Darwin was working on the topic of the origin of species, which is why Wallace sent his 1858 paper to Darwin. At this point we can imagine that publication of the 1844 essay—with Hooker's and, to a lesser extent, Lyell's endorsement—would have had a significant effect. In these circumstances, Wallace might have become aware of the selection theory before writing his 1858 paper (unless, being in the Far East, he missed the publication of Darwin's essay).

These counterfactuals alert us to the point that in a world entirely without Darwin, there would have been no one trying to promote evolutionism to figures such as Hooker, Lyell, and Gray in the 1850s. There would be no behind-the-scenes promotion of the selection theory, and the world would also be missing a major source of support for the more general idea of divergent, adaptive evolution. We have to be sure that we can imagine the other naturalists moving in this direction without Darwin's prompting. It is worth remembering that there were evolutionary theories that did not focus on local adaptation, and in the alternative universe such non-Darwinian theories may have played a greater role than they did in our world. It must also be acknowledged that widespread recognition that a respected figure such as Darwin was working on the issue may have helped create a more tolerant attitude toward evolutionary theorizing in the scientific community. Making the case for the emergence of an evolutionary viewpoint in a world without Darwin is a little more demanding than we might initially think. On the other hand, since most naturalists had only the vaguest idea of where Darwin was heading, we can probably afford to discount this indirect influence, provided that we can find sufficient evidence for independent developments pointing in a similar direction.

The worldview of the early nineteenth century had been transformed in one crucial respect, without which evolutionism would have been literally inconceivable. In the previous century, it had been widely believed that the earth was only a few thousand years old, so the Genesis account of creation could be taken more or less literally. As late as 1802, William Paley's *Natural Theology* still assumed that there had been only a single creation in which the modern species had been formed. But this assumption came under threat when geologists established an outline of the earth's history in which huge changes had taken place over vast periods of time.

The biblical vision of the past left no room for a period of prehistory predating the appearance of the first humans. But fossil evidence proved that the earth's inhabitants had changed over the course of geological time. There was much debate about how much time was involved, but all geologists recognized that earth history occupied a period of time vastly more extensive than human history. Darwin favored Lyell's "uniformitarian" approach, which postulated almost indefinite amounts of time and dismissed as an illusion the apparent discontinuities separating the geological periods. As fossil discoveries poured in, they tended to fill in any gaps, making the possibility of some form of continuity between successive populations more plausible. Most important of all, the fossil record seemed to indicate a progressive trend in the history of life, which Lyell alone tried to discount. The implication that the development of life on earth represented an unfolding toward the modern state of affairs chimed with growing enthusiasm for the idea of progress in social history.[4]

The establishment of the modern vision of earth history was one of the great scientific triumphs of the period from 1790 to 1840. By the latter date, the sequence of geological periods that we recognize today had been established and an outline of the history of life had emerged from the fossil record. Most geologists thought that major changes in the earth's crust, such as the formation of mountain ranges, were brought about by catastrophic upheavals. William Whewell coined the term "catastrophism" to denote their position. Until the early 1830s, most scientists assumed that the last of these catastrophic events could be identified with Noah's flood, although by 1840, people had abandoned the notion of a geologically recent universal deluge. But catastrophists continued to see upheavals as punctuation marks separating the geological periods and their different fossil populations. They assumed that the extinction of species was a consequence of these disasters, with new species then being introduced to fit the different conditions of the next period. The question of where the new species came from could hardly be avoided. In Britain, most catastrophists assumed events of miraculous creation, while on the continent it was less fashionable to invoke divine interference. Evolution became a possibility if it was accepted that living things could survive the major transitions, although evolution was not the only mechanism under consideration.

Catastrophists recognized that although their hypothetical earth movements were very rapid, erosion and deposition required vast periods of time to form layers of sedimentary rock during each geological period. Theirs was not a young-earth chronology, although the timescale they envisaged was much shorter than the one that we accept today. In the middle decades of the century, there was a fairly broad consensus that the earth was around a hundred million years old. The main challenge to this consensus came from the uniformitarian geological theory proposed in Charles Lyell's *Principles of Geology* (1830–33). Lyell wanted to free scientific geology from any link to the biblical story of creation. He didn't like catastrophes because the last one was still, at that point, identified with the Genesis flood. He was also suspicious of the catastrophists' assumption that there was a starting point for the whole geological sequence, which they took to be evidence of the planet's creation. He proposed an endless cycle of gradual, natural changes spread over an unimaginably vast period of time, of which we have evidence only of the later stages. Mountains were built up over millions of years by earthquakes no bigger than those we observe, and all erosion is the product of the natural flow of streams and rivers on the same extended timescale. Contrary to popular belief today, Lyell was not successful in converting the catastrophists to his point of view. Darwin was his most enthusiastic disciple, and Lyell's views might not have gained support as rapidly as they did without Darwin's efforts. But what Lyell did eventually achieve, and would almost certainly have achieved anyway, was a greater recognition of the power of natural agencies to transform the earth, and a scaling down (if not an elimination) of the hypothetical catastrophes. By the middle of the century, few geologists still believed that each catastrophe wiped out all life on earth—there was at least some continuity between the successive periods.

The need for an explanation of how new species appeared in the fossil sequence might have seemed obvious. But most naturalists were reluctant to speculate explicitly—it is almost as though they preferred to put off thinking about the issue. The theory of successive miraculous creations was an obvious compromise, although few wanted to imagine the details of how a supernatural intervention might work. Instead, naturalists looked for patterns in the fossil record that might indicate an overarching divine plan. There was an obvious element of progress in the fossil record, from primitive invertebrates through the ages of fishes, reptiles, and mammals, concluding with the appearance of humans. Perhaps more advanced forms of life had been created as the earth's physical environment improved over the course of geological time. Or perhaps the Creator had deliberately symbolized humanity's key position in His plan by imposing a progressive pattern on the animal kingdom with the human form as its goal. The anatomist Richard Owen made the suggestion in the 1850s that in each new class, lines of specialization radiated out toward the more advanced modern members. Although reluctant to condone transmutation openly, it did not escape his attention that some of the patterns he was seeing in the fossil record looked remarkably like the sequence an evolutionist would expect to find.[5]

THE ORIGINS OF TRANSFORMISM

The possibility that new species might arise from the transmutation of earlier ones had been raised in the closing years of the previous century. (The term "evolution" did not become common in its modern context until later. "Transmutation" was the popular term in the early nineteenth century, although Darwin preferred "descent with modification.") The transformation of existing forms by natural processes had been suggested by radical thinkers in the eighteenth century, although modern scholars tend to be wary of seeing too close a resemblance to the modern theory of evolution. By the 1790s, fully fledged theories of organic development had emerged. One pioneer in this field was Darwin's grandfather, Erasmus

Darwin, a noted medical doctor and poet, whose *Zoonomia* of 1794–96 envisaged living things struggling to improve themselves over vast periods of time, gradually advancing life toward ever more complex structures. Darwin even incorporated this vision into his poetry, which ensured it some influence outside the scientific and medical communities. Paley's *Natural Theology* was written in part to defend the argument from design against Darwin's claim that adaptation was a natural process resulting from the purposeful activities of living things. The young Charles Darwin was aware of his grandfather's speculations and shared his belief that sexual reproduction was the key to understanding nature's expansive powers. Erasmus Darwin also appealed to an element of struggle in nature, but he saw this in terms of predators and prey, not competition between the individuals within a population.[6]

A far more potent figure for those seeking a radical alternative to divine creation was the French biologist Jean-Baptiste Lamarck. The theory he outlined in his *Zoological Philosophy* of 1809 has been remembered because one element of it—the inheritance of acquired characteristics, often known as "Lamarckism"—emerged as the most plausible alternative to natural selection in the late nineteenth century. It is probably true to say than Lamarckism and natural selection are the only two mechanisms of *adaptive* evolution that have ever been suggested. This is why neo-Darwinism emerged as the dominant theory of evolution once Mendelian genetics showed that acquired characters cannot be passed on to future generations. Curiously, in view of its later reputation, Lamarckism was only a secondary component of Lamarck's own theory. He thought that there was a more fundamental progressive trend driving species up the scale of complexity. Adaptive modifications generated deviations from what would otherwise be a linear chain of being from the simplest organisms to humans. Lamarck and his early followers seem to have believed in multiple parallel lines of progressive evolution. Although he did provide a diagram of what looks like a branching tree of relationships, most historians think that Lamarck did not really anticipate the theory of common descent.

Older histories of biology assumed that Lamarck was marginalized within French science by his archrival, Georges Cuvier. But recent studies show that there were also radical French naturalists who took a more sympathetic view of Lamarck's theory. During his early years studying in Edinburgh, Darwin heard about Lamarck's theory from the anatomist

Robert Grant, and Adrian Desmond has shown that there was an active campaign to challenge the argument from design by radicals within the British medical profession. At the time, there was much opposition from the scientific and medical establishments, with Richard Owen emerging as the leading scourge of the transmutationists in the 1830s. Owen did his best to undermine the case for the continuous development of life on earth, pointing to the many gaps in both the fossil record and the arrangement of living species. Only later did he try to modernize the argument from design to incorporate what looks suspiciously like a form of evolution.[7]

The radicals also became excited about a new vision of how species were related called "transcendental anatomy," which looked to similarities beneath superficial adaptive functions to show that there was a fundamental unity in the animal kingdom. The transcendental approach was pioneered in Germany and brought to Britain by Owen, who persuaded natural theologians that the unity was itself an indication that nature followed a divine plan. But transcendentalism was not the only expression of a more broadly based structuralist vision of nature. Many were suspicious of the woolly minded idealism of writers such as Lorenz Oken, with their talk of structural patterns embedded in the very fabric of the world. Followers of J. F. Blumenbach looked to natural processes as alternatives to the biblical view of creation, although their ideas included processes besides transmutation. In France, another of Cuvier's great rivals, Geoffroy Saint-Hilaire, promoted a more materialistic version of the "unity of nature" thesis. Geoffroy argued that the similarities uncovered by the comparative anatomist might indicate that one species was derived from another. He proposed a saltationist or macromutationist version of transformism in which the process of individual development was occasionally distorted so that an embryo matured into a form significantly different from its parents. Perhaps a change in the external environment was supposed to have triggered the transformation, although Geoffroy doesn't seem to have believed that the new form was necessarily better adapted to the new conditions. He used his theory to explain how modern crocodiles could have appeared as modified descendants of earlier crocodilians whose remains had been found in the fossil record.[8]

The claim that the first individuals of a new species might emerge suddenly, as dramatically transformed versions of their parents, has been ridiculed as the theory of the "hopeful monster." The monster in such a case would have to be lucky, because a drastic interference with the process of individual development would offer no guarantee that it would be adapted to the environment or even viable (most severe mutations are, in fact, fatal). To found another species, it would then need to find a similar monster of the opposite sex—an unlikely proposition, given that saltations of any kind tend to be rare. Saltationism is very different from modern Darwinian evolutionism, and it allows the naturalist to retain a belief in the fixity of species, because the parent form remains unchanged when it produces mutated offspring. But the idea does at least involve transmutation and could provide the basis for a theory of common descent. As long as the transformations were not too drastic, one could still recognize which new species were the products of mutations within a particular common ancestor.

RIVAL THEORIES

Geoffroy's saltationism shows that in their search for an alternative to divine creation, naturalists conceived versions of the transmutation theory that looked nothing like the Darwinian model of slow, adaptive change. There were other alternatives even less like the modern view of evolution. In a letter to the publisher John Chapman in 1848, Richard Owen claimed that he could think of a half dozen natural mechanisms by which new species could be introduced. Lamarckism and saltationism were two obvious candidates, but what were the others?[9]

One was the idea that a new species might appear as a result of hybridization between two existing forms. Distinct species were supposed to be infertile, but it was widely suspected that they might occasionally be able to interbreed successfully, producing a hybrid combining characteristics from both parents. If this bred true, it would constitute a new species. The process was thought to occur most readily in plants, and modern biologists recognize that new species have sometimes been produced in this way. In the mid-eighteenth century, the great systematist Carolus Linnaeus suggested that God might have created only a single founder type for each genus. The groups of related species we now find in each genus were formed over the course of time by acts of hybridization between the original parent forms. The total number of species in the world would thus increase over time by purely natural means. The possibility that new species might be produced in this way was taken seriously throughout the nineteenth century. Some historians of genetics believe that the hybridization experiments undertaken by Gregor Mendel in the 1860s—which we take as the foundation of genetics—were actually inspired by the hope of promoting the hybridization theory as an alternative to Darwinism. One geneticist, J. P. Lotsy, argued that hybridization produced new characteristics and was thus the source of all new species. He even suggested that vertebrates might have originated through the interbreeding of two invertebrate types.[10]

Owen was also fascinated by the phenomena of metamorphosis and the alternation of generations. We are all familiar with insect species that pass through different phases during their life cycle, as in the case of the caterpillar and the butterfly. Some invertebrate species exhibit even more dramatic changes—they breed in one form for a number of generations and then switch to an entirely different form. Owen may have wondered what would happen if these phases became permanently separate, for instance, if the caterpillar became sexually mature and began to reproduce without needing to turn into the adult form of the butterfly. We would then see two entirely distinct species, never realizing that they had once been merely separate phases in the life cycle of a single species.

The problem with this idea, as with the hybridization theory, is that it doesn't explain the origin of the original forms that differentiate over the course of time. Linnaeus simply assumed that the original forms of each genus were divinely created. Owen may have preferred not to think about this question, but he must have realized that, unlike any form of transmutation theory, neither of these alternatives could offer a meaningful explanation of the ultimate origin of the diversity of living forms. There might be a naturalistic multiplication of the number of distinct species, but all the basic characters must already be in existence for the process to begin. Lotsy evaded this problem by suggesting that hybridization actually produces new genetic material, but his theory would still require the prior existence of at least some distinct original forms.

There was another possibility that avoided this problem, although it seems quite bizarre to us today. In the eighteenth century, radical thinkers suggested that unstructured organic matter might in some circumstances be able to organize itself into living creatures, and some experiments seemed to indicate that microorganisms could come about by such a process of "spontaneous generation." Lazzaro Spallanzani subsequently discredited these experiments, although the outcome of this debate was not as clear as modern textbooks pretend.[11] Both Lamarck and Erasmus Darwin continued to believe that the process actually occurred, thus explaining the primitive origins from which the evolutionary progress began. But in the previous century, the naturalist Buffon had wondered if spontaneous generation might, under very rare conditions, produce more complex living structures directly from unorganized matter. This would avoid the need to invoke a process of transmutation.

Historians have long assumed that this idea fell out of fashion around 1800, leaving only the more limited notion of the spontaneous generation of very primitive forms as an alternative (though a highly controversial one). But Nicolaas Rupke has shown that an equivalent idea that he calls "autochthonous generation" survived well into the nineteenth century. The German anatomist Karl Vogt and the Austrian botanist Franz Unger both accepted it, and eminent thinkers including Alexander von Humboldt and Johannes Müller also considered it.[12] It was another manifestation of the formalist tradition, because it assumed that the fundamental laws of nature predetermined new forms of life. By supposing that new forms could be produced directly from matter, this early version of formalism deflected attention away from transmutation. But most formalists conceded that there would also be predetermined evolution within the major organic types, and as the century progressed, they gradually increased this element of their theory at the expense of autochthonous generation.

Historians lost sight of autochthonous generation because it fell rapidly out of favor in the middle decades of the century. Unger abandoned the theory in the 1850s, realizing that the new theory of the cellular basis of life made it implausible to think of multiple-celled organisms being spontaneously generated. He converted to a form of evolutionism, and Vogt too went on to become an enthusiastic evolutionist. This episode tells us a great deal about the complex state of naturalists' thinking in the early decades of the nineteenth century. It shows that, just because few naturalists openly favored transmutation, we cannot simply assume a blanket support for divine creation. There were many who had serious doubts about the idea of multiple supernatural events throughout the earth's history. But it was by no means clear at first that transmutation was the best alternative, so it was better to hedge one's bets and say nothing in public—the strategy that Owen adopted.

As early as 1836, Lyell wrote to the highly respected astronomer and philosopher of science Sir John Herschel in terms suggesting that they both thought it probable that the origin of new life forms was the result of "intermediate causes," meaning a process governed by natural law. Yet Lyell and Herschel, like Owen, still thought these laws would somehow embody the Creator's forethought. Charles Babbage's unofficial *Ninth Bridgewater Treatise* of 1838 made it clear that design—including the design of new species—was best imagined working though law rather than miracle. All of these eminent figures were content to let it be assumed that they favored some form of "creation by law"—a loosely defined concept that allowed the Creator to impose his designs on the world through processes governed by the laws he had instituted. Exactly what those processes were, it was best not to inquire.[13]

The most obvious reason to be cautious was fear of public reaction, but another was the sheer number of possibilities to be considered. The few radical thinkers who did openly postulate naturalistic explanations of the origin of species had to explore a much wider range of alternatives than we might imagine. Transmutation was only one possibility—and a dangerous one, given the negative reaction of conservative scientists such as Cuvier to Lamarckism. But in the 1840s and 1850s, open support for miracles decreased as more thinkers began to extend the role of natural law into this area. The range of alternatives also seems to have narrowed as the most bizarre ideas were increasingly recognized as such, allowing attention to focus on transmutationism as the best way forward. Even without Darwin's input, naturalists and other concerned thinkers were beginning to adopt a more positive attitude toward what we call evolution.

VESTIGES AND AFTER

The most obvious indication of this changing attitude is the reaction to a book published in 1844, fifteen years before the *Origin of Species*. Entitled *Vestiges of the Natural History of Creation*, it was written by the Edinburgh publisher Robert Chambers, but published anonymously. Chambers was an amateur naturalist, and he thought that the specialists who dominated the scientific community were reluctant to admit that new discoveries were pointing toward a worldview based on natural development rather than miracles. He also had a political agenda. As a leading publisher of material aimed at the rising middle class, he wanted to convince his readers that social progress was inevitable. One way to do this was to show them that human history was merely a continuation of a universal process of development. In the words of historian James Secord, this was a "popular science of progress." To make such a radical idea palatable to his middle-class readers, Chambers suggested that the laws governing development—physical, organic, and social—were instituted by the Creator as a means of achieving His ends indirectly rather than through miraculous intervention.[14]

Vestiges presented a vision of universal history governed by natural law. It postulated that the planets were formed as a primeval nebula of rotating dust collapsed under its own gravity. This "nebular hypothesis," pioneered by the French astronomer Pierre-Simon Laplace, had been widely promoted as a model for understanding how the physical world could be brought into its present state by the operation of physical laws.[15] It thus provided a way of encouraging people to think in evolutionary terms. But its implications were strikingly non-Darwinian, since the process moved inexorably in a single, progressive direction: from dust cloud to an orderly system of planets. For Chambers, it served as the starting point for a more general evolutionary cosmology based on the inevitability of law-bound progress.

Once the earth was formed, Chambers supposed that electricity generated simple living forms from inorganic material. Successive generations of living things became gradually more complex, leading eventually to humankind. Here was the whole materialist program of what we call evolution, as sketched out earlier by Lamarck and Erasmus Darwin, revived and presented once again to the public. But Chambers had little interest in adaptation, and his model of development was not based on the tree of life. He saw multiple parallel lines of evolution all moving in similar directions, each driven by the inbuilt tendency to progress. Evolution was a predetermined process much like the development of the embryo toward maturity. There was hardly a naturalistic explanation of change because it was based on the unfolding of a program built into the laws of development. The term "program" is only partly anachronistic here, since Chambers used an argument pioneered by the inventor of the computer, Charles Babbage, who suggested that God could program the universe to change just as Babbage could predetermine the operations of his mechanical computer. All was based on law rather than miracle, and humans were the inevitable outcome of this law-bound process of development.

Conservative scientists reacted to *Vestiges* with horror—this was the worst kind of materialism thinly disguised by the claim that the Creator instituted these laws. Darwin himself was forced to think deeply about his new theory when he read the hostile review of *Vestiges* written by his Cambridge geology mentor, Adam Sedgwick. Historians have traditionally assumed that the book had little impact on science, but James Secord's detailed study of the public reaction to it shows that responses varied according to location and social class and—more importantly—changed significantly over the following decade.[16] In the end, many ordinary people did begin to think seriously in terms of a naturalistic model of development. Lord Alfred Tennyson's *In Memoriam* is just one example of a literary work inspired in part by this vision and seeking to come to terms with its implications. Benjamin Disraeli parodied popular enthusiasm for the development hypothesis in his 1847 novel, *Tancred*. These literary allusions also remind us that attitudes were changing on a wider front. Fewer educated people now felt comfortable with the traditional Christian view that all humans were damned unless they accepted the Bible's offer of salvation through Christ. Nor did people invariably see the world of natural law as a harmoniously balanced utopia designed by a benevolent God; in Tennyson's words, the world of life was "red in tooth and claw." Without any input from Darwin, people were beginning to see the universe as a scene of progress driven by struggle, and they were willing to accept that humanity was the outcome of such a process. Social progress was simply an extension of universal law into a new realm that might well operate by the same harsh rules.

What of the scientific community? In Britain, the older generation was almost universally conservative and hostile, but attitudes among younger, more radical naturalists were mixed. A good indication of the ensuing tensions can be seen in the work of Richard Owen, who had once savaged the Lamarckism of radicals such as Grant. Owen refused to write a hostile review of *Vestiges* and openly implied, in the conclusion of his *On the Nature of Limbs* (1849), that the development of life on earth might unfold by divinely implanted laws. He went on to stress that the fossil record indicated the divergence of specialized descendants from a generalized ancestor, a line of evidence Darwin seized upon to back his branching-tree model in opposition to simpleminded progressionism. Owen's conservative backers seem to have discouraged his efforts to move toward an evolutionary perspective in the 1850s, but in the aftermath of the *Origin* debate, he was quite happy to counter Darwinism with the suggestion that evolution unfolded in accordance with divinely implanted laws. It is hard to believe that he would have held off indefinitely from making the same move in a world without Darwin.[17]

One might have expected an aspiring professional like Huxley to favor the theory, but in fact he wrote a vitriolic review of a later edition of *Vestiges*, pointing out that Chambers had fudged the fossil evidence to make the progress of life seem more continuous than the record indicated. More seriously, Huxley recognized that the book offered only a vague ideology rather than a workable scientific theory. Its implication that progress was built in according to the Creator's plans was really a device for preserving the old argument from design. This was exactly the compromise that Owen was working toward and that Huxley distrusted. But where Huxley suspended judgment, other younger naturalists—most obviously Wallace—were inspired by *Vestiges* to think more openly in terms of natural laws governing the emergence of new species. Eventually Huxley too moved in the same direction—prompted by Darwin, of course, but there were other pressures that might have forced him in the same direction without Darwin. Provided that the element of design could be eliminated, evolutionism became a weapon to use against natural theology. This would play into the hands of rising young professionals like Huxley, anxious to present science as an alternative to religion in national affairs.

At the end of the 1850s, Chambers suggested to his publisher, John Churchill, that he should offer a prize for the best essay on the scientific credentials of the development hypothesis.[18] The competition did not take place, but there were by now several influential figures who Chambers hoped would rise to the bait. They included the physiologist William Benjamin Carpenter, the science writer George Henry Lewes, the mathematician and philosopher Baden Powell, and the social philosopher Herbert Spencer. The latter had come out openly in favor of Lamarck's theory in 1851 and had published an evolutionary explanation of how the human mind had developed in his *Principles of Psychology* of 1855. Spencer noted that his technique of explaining the origin of the mental faculties in terms of how they have been used in everyday life implied tacit acceptance of the "development hypothesis"—what we call the theory of evolution.[19]

In the same year, Powell's *Essays on the Inductive Philosophy* argued that the Creator's powers were revealed more clearly in the effects of the laws He imposed than in miraculous interventions, with the clear implication that this applied to the history of life on earth.[20]

There were still ideological divisions between the influential figures that Chambers hoped to inspire. Powell was a liberal Anglican who would feel more comfortable with the idea that the laws governing evolution were the expression of a divine plan. This was also the position that Owen favored. Spencer and Huxley were active opponents of organized religion who wanted evolution to be based on natural laws understood in the everyday sense of observable regularities in the sequence of events. Even without the selection theory, evolutionism would have served this purpose, provided a natural process such as Lamarckism or saltationism could be invoked. For Huxley especially, the move to naturalistic evolutionism would be exploited as a valuable weapon in promoting the interests of the small but increasing number of professional scientists.

Parallel moves were afoot in other countries, especially in Germany. Radical figures such as Karl Vogt were switching their attention from autochthonous generation to transmutation. More seriously the palaeontologist Heinrich Georg Bronn produced an essay in 1858 that explored the patterns of development shown by the fossil record and used the image of a branching tree to represent the overall effect. Although not a convert to evolutionism, Bronn illustrates a more sophisticated approach to the issues that went far beyond both creationism and the simple model of progress designed to reflect the developing human embryo. In our world it was Bronn who translated the *Origin of Species* into German (with some notable modifications that obscured the more radical implications of the selection theory). This translation provided the inspiration for perhaps the most robust advocate of Darwinism in Germany, Ernst Haeckel. But by the time he read Darwin, Haeckel was already beginning to think in terms of a naturalistic theory of development, and his Darwinism was always tinged with significant elements of Lamarckism and German transcendentalism (Darwin, Lamarck, and Goethe were the three luminaries of his evolutionary pantheon). As with Owen, it is difficult to imagine that in a world without Darwin, Haeckel would have worked through the 1860s without coming to the conviction that natural processes of development controlled the history of life on earth.[21]

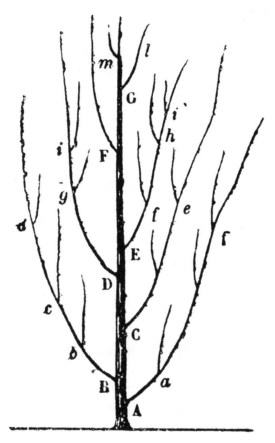

Figure 4 · H. G. Bronn's illustration of the development of life, from page 907 of his 1861 "Essai" (submitted in response to the Paris Academy of Science's 1850 call for an analysis of the laws of distribution in the fossil record). Note that although the illustration portrays a treelike pattern of divergence, it still includes a main trunk leading directly upward, presumably toward humankind.

CHANGING VALUES

These debates in the scientific community were paralleled by developments in Western culture as a whole. Chambers had aimed his *Vestiges* at middle-class readers in order to convince them that they could preserve some of their traditional values while moving toward a new ideology of social progress. James Secord's survey of the public reaction reveals the extent to which values changed between 1844 and 1859. We have seen these changes in the scientists, but have also noted how writers such as Disraeli and Tennyson, in tune with both scientific discoveries and the new values, responded. Ordinary people no longer believed that the Bible story of creation could be taken literally. They were increasingly less willing to accept the traditional Christian vision of original sin and redemption solely through acceptance of Christ's sacrifice. As Darwin himself noted, this would mean that most of his family and friends were damned to eternal torment in hell. People wanted a new view of God's relationship to humanity that made room for individual moral activity and effort. Chambers showed that it now made better sense to see human social progress as a continuation of the development of life on earth, both processes governed by natural law rather than by divine miracle. Charles Lyell noted that the popularity of *Vestiges* stemmed from a growing feeling that an indefinite series of miracles through earth history now seemed implausible. The problem was that this change of values forced everyone to confront the prospect that primitive humans had emerged from an animal, presumably from an ape ancestry.[22]

It was Herbert Spencer who turned Chambers's vague idea of a law of progress into an evolutionism in which progress was the inevitable outcome of individual effort and achievement. In a changing environment (physical, economic, or cultural), individuals adapt their behavior to improve their prospects—and if the Lamarckian effect were valid, the resulting improvements would be transmitted to future generations. Spencer was an enthusiast for free-enterprise liberalism, quite willing to see competition as a factor that would increase the pressure on individuals to adapt. His *Social Statics* of 1851 advocated an extreme form of laissez-faire economics that accepted that those who failed to rise to the challenge would suffer. In our world, Spencer subsequently became known as an advocate of social Darwinism, not least because he coined the phrase "survival of the fittest." But long before Darwin, and thinking along Lamarckian rather than selectionist lines, he had recognized that struggle might be the driving force of progress in both the animal and the human worlds.[23] This warns us not to jump to conclusions when it comes to assessing Darwin's cultural impact. Many aspects of what we call the "Darwinian" worldview were already in place before he published, which is why left-wing critics of free enterprise have always claimed that Darwin merely projected the social values of his time onto nature.

An important component of the new attitude can be traced to the influence of Malthus's principle of population. There had been a huge debate over the implications of Malthus's book in the early nineteenth century, with many condemning the harshness of his suggestion that the poor should be left to suffer the consequences of their improvidence. Natural theology had always recognized some aspects of competition, most obviously the predator-prey relationship. Erasmus Darwin had expounded the superfecundity of nature, but assumed that the outcome of the resulting struggle was beneficial in the end. Malthus made the same assumption, but many thought his principle had much darker implications.

Malthus did not explicitly translate his principle into an ideology of cut-throat competition. Indeed, he coined the term "struggle for existence" when writing about warfare between tribes rather than competing individuals. But the idea that population pressure imposed restrictions on resources that would entail individual competition gradually became more widely appreciated. Spencer himself accepted this view, although he thought population pressure would eventually diminish as we became more intelligent (and hence, in his rather naive view, less sexually active). Surveys of the literature of the period show that it was frequently accepted that social interactions would be based on the desire for personal advancement, with little sympathy for those who could not hold their own against rivals. Charity was reserved only for those who had suffered misfortune through no fault of their own. As this dog-eat-dog attitude insinuated itself into people's behavior, Spencer's writings encouraged everyone to think that it was just a natural application of the way nature itself operated. Darwin may have highlighted the harsh implications of this image of nature, but he certainly did not originate it.[24]

Not everyone found the new values easy to accept, and Tennyson's *In Memoriam* is a good illustration of the resulting tensions. Devastated by the loss of a close friend, Tennyson took refuge in the belief that death itself was a necessary part of the process of cosmic progress. This personal tragedy made it easier for him to appreciate the harsher image of nature that was now emerging both in science and society. The fossil record showed him that nature cared nothing for species—it was full of bizarre forms that had gone extinct over the course of time. Lyell's uniformitarianism implied that extinction was a natural process, and by this time naturalists were becoming aware that human activity was causing the extinction of some species. Tennyson, who famously portrayed nature as "red in tooth and claw," a scene of constant struggle between predators and prey and between rivals within the same population, was not alone in recognizing that it was no longer quite so easy to accept the natural theologians' vision of a harmonious world.[25] Critics see Tennyson's poem as expressive of values consonant with Darwinism long before Darwin published, which reveals that those values were already widely available. In fact, Tennyson did not think in terms of natural selection, then or later—he did not anticipate Darwin's creative extension of the struggle metaphor. His vision shows us that it was quite possible to think in terms of a harsh nature that consigned her less successful products to the scrapheap without appreciating the logic of individualistic natural selection.

In our world, Darwin's decision to begin writing his big book on natural selection in the mid-1850s was influenced by a growing sense that the climate of opinion had changed to make such a hypothesis more acceptable to both the public and the scientific community. We have now seen that almost everyone else understood the situation in the same way. The pressure for change was building up, and even without Darwin's intervention, the 1860s would almost certainly have seen more naturalists arguing openly for an evolutionary perspective. Wallace, Hooker, and Gray would have explored the biogeographical evidence, which would chime with the branching model of development being explored by Owen and Bronn. More radical thinkers such as Spencer and Haeckel would almost certainly have wanted to reopen the case for Lamarckism as an explanation of adaptive change within a progressionist worldview. Even Huxley would then have been forced to reconsider his distrust of the development hypothesis—despite his enthusiasm for the *Origin*, he did not use evolutionism in his own scientific work until after he had read Haeckel.

A number of other factors unconnected with Darwin also came into play around 1860, each helping to focus attention on the topic of evolution. Huxley and Owen clashed on the relationship of apes to humans, while archaeologists and anthropologists pioneered a new model of human prehistory. Arguments that had died down since the publication of *Vestiges* were now being reopened, and this would have happened even without the appearance of the *Origin of Species*.

4

THE EMERGENCE OF EVOLUTIONISM

The publication of the *Origin of Species* is widely seen as the defining moment of the Darwinian revolution. Even at the time, it was apparent to many readers that the book marked a turning point. Darwin offered a wealth of evidence in favor of the theory of common descent, and a radically new hypothesis about the mechanism of change. He wrote in a style that was accessible to ordinary readers, and he called in examples from areas familiar to all, including natural history, gardening, and animal breeding. The story of its impact has, if anything, grown over the decades, swelling to a crescendo in 2009, which makes it difficult for the modern reader to gain a true perspective.

The *Origin* was important, but much of its influence was indirect. Opinions differ on the effectiveness of Darwin's presentation: the evidence may have evoked familiar images, but his writing style was seldom inspirational. Literary scholars such as Gillian Beer and George Levine praise the book's effectiveness, but it did not become a bestseller until cheap reprints became available long after it was recognized as a classic.[1] The first edition sold for fifteen shillings, putting it out of reach of all but the well off, and sales in Darwin's lifetime were much lower than those for equivalent writers such as Chambers and Spencer. During the crucial years of debate, most people got their information about Darwinism secondhand, so we need to think very carefully about the book's direct impact. There were many commentaries about the volume, both favorable and critical, so other writers must have played a major role in disseminating the theory. The debate hung in the balance for several years, with the general theory of evolution only beginning to seem secure by the late 1860s—by which time Spencer and others would have entered the scene anyway. As it was, natural selection was still widely rejected. Our assessment of the book's success may be clouded by the way in which Darwin quite rapidly became the symbol or figurehead of the evolutionary movement, a position he has retained to this day. Is it possible that his influence was boosted by factors supplementary to his scientific arguments? And if so, does this allow us to imagine other inputs that may, in a less dramatic way, have forced everyone to start taking the idea of evolution seriously?

Two factors that boosted Darwin's image were his name and his appearance. They may seem trivial, but historians are now very much aware of the roles played by rhetoric and presentation in the promotion of a new idea. It was T. H. Huxley who coined the term "Darwinism" in his 1860 review of the *Origin*, and the term rapidly achieved wide currency, along with the adjective "Darwinian."[2] The name seems to trip off the tongue, very much in contrast to potential alternatives such as "Wallaceism" or "Spencerism." Herbert Spencer did, at least, see the term "Spencerian" coined as an adjective for his evolutionary philosophy, but the noun "Spencerianism" did not come into general use. The ease with which his name could be used as a label certainly helped elevate Darwin to the level of a figurehead for the evolutionary movement.

Darwin biographer Janet Browne has explored the second image-boosting factor.[3] During a bout of illness in the early 1860s, Darwin let his beard grow, transforming his appearance into that of a venerable sage. By the end of that decade, photographs, portraits, and caricatures came into wide circulation. Even in France, where Darwinism was not very popular, depictions of a bearded Darwin alongside a monkey began to appear in cartoons to symbolize the implication of the new model of human ancestry. The same image of Darwin still served as an icon in the 2009 celebrations. Here, as with the popularity of the term "Darwinism," it is worth thinking about the extent to which such presentational factors boosted the impact of the *Origin of Species* in ways that transcended the mere force of its argument. These factors have also kept Darwin in the forefront of public attention to this day. Yet historians are increasingly certain that Darwinism was not the only source of evolutionary thinking. Spencer's philosophy was probably far more influential than the theory of natural selection, especially outside science (though its impact on scientists should not be dismissed). Yet Spencer's image adorned few cartoons; photographs reveal a somewhat unappealing visage that seems to have sparked little inspiration among cari-

Figure 5 · Caricature of Darwin from *Punch*, 1881. The worm relates to the topic of his last book, but the caption reminds us that the real issue was humanity's descent from the apes.

caturists. Today Spencer's philosophy has been forgotten, while Darwin's theory has gone from strength to strength, making it all the more difficult to gain an accurate perspective on their relative influence at the time. Wallace, to be fair, did acquire a beard as resplendent as Darwin's own, but as his modern defenders complain, he was something of an outsider, then as now.

None of this is meant to detract from Darwin's reputation. His book would have had an immense impact even if his name and appearance had been less memorable. But we need to be aware that the promotion of a new theory is not just a matter of presenting ideas and evidence that speak for themselves. It is a social process in which presentational factors do play a role. Taking Darwin out of the picture removes not only his theory and his arguments, but also the symbolic role he was able to play as a figurehead for the new way of thinking. The fact that he served as a figurehead should force us to think more carefully about his impact, making it easier to contemplate the possibility that other influences might have brought about the transition to an evolutionary worldview.

Without Darwin the transition to evolutionism would have been more gradual and less traumatic because there would not have been a single event to trigger the changeover, and none of the new inspirations would have seemed as threatening to traditional values. By the late 1860s, both the scientific community and the general public would have begun to accept a general evolutionary perspective, lacking only the concept of natural selection acting on individual variation. There would be a gradual accumulation of smaller stimuli, not a single focus of attention. Those involved may not even have imagined themselves to be living through a scientific revolution. There would still have been debates over topics such as the implications of an animal ancestry for humankind and presumably some tension. But there would be no single figure who could have been hailed as the instigator of a materialist revolution, and the historians of this alternative universe might have a more realistic perspective on how the transformation took place.

One way of trying to understand what might have happened is through a comparative international study of how evolutionism developed in the late nineteenth century. Darwinism was not received in the same way throughout the world. Britain, America, and Germany experienced a full-scale Darwinian revolution in the early 1860s—indeed some Germans thought their country had become the true home of Darwinism. Whether the German *Darwinismus* was the same thing as British or American Darwinism is open to doubt, though, since some aspects of Darwin's theory were quite difficult to translate and German biologists had their own indigenous approach to the topic. Elsewhere the impact of the *Origin of Species* was less acute. In parts of the world where there was no strong scientific tradition, evolutionism gained ground almost exclusively in the context of calls for social progress, and it was Spencer, not Darwin, who became the leading influence. Even in America, Spencer's evolutionary ideology shaped much of the discourse in the later nineteenth century. In France, the *Origin* had only limited impact, partly because the first translation took a distinctly anticlerical bias, but also because French naturalists found it hard to take natural selection seriously. Darwin gained an iconic status as a symbol of evolutionism, but the general idea of evolution was only slowly taken up by the scientific community, and then mostly in the form of a revived Lamarckism. In this case the source was not Spencer but the residue of an earlier generation of French theorizers including Lamarck himself.[4]

The French experience offers a model through which we can understand how events might have unfolded in Britain, America, and Germany if we imagine Darwin's book taken out of the equation. In the absence of the selection theory, other factors would have prompted a more gradual acceptance of evolutionism based on alternative mechanisms such as Lamarckism. In Britain and America, Spencer would have been influential, while the Germans would have looked more to Continental sources of inspiration, including both Lamarck and their native traditions of formalism and transcendental philosophy. There would have been a stronger input from traditions that did not assume that adaptation was the sole determining factor shaping the structure of living things. Evolutionism would have emerged gradually into scientific and public discourse, securing itself by the 1870s. This might be a few years later than in our own world, because there would not have been the sudden stimulus provided by the *Origin*, but the transition would have aroused less antagonism from conservative thinkers.

Whether it would be perceived as a revolution is hard to determine; possibly it would seem more like a natural transition, the culmination of a broad trend in nineteenth-century thought. There would almost certainly be no single figure who would take on Darwin's iconic status. Evolutionism would be seen as a group effort, with a number of important contributors each having something to offer toward the creation of the new perspective. The group would include most of the figures we are familiar with, although the relative significance of their contributions might be different. Spencer and Haeckel would be key players, but I am not quite so sure about Huxley.

Some of the wider trends in thinking were simple continuations of factors already active during the 1850s, including the growing distrust of miracles and the enthusiasm for the idea of progress. Others were new and potentially more revolutionary in nature. There was a debate over the relation between humans and apes triggered by new information about the great apes in Africa. Recognition that there was serious evidence for a long period of prehistory during which humans had lived in the Stone Age dramatically transformed archaeologists' thinking, with major implications for the evolutionary perspective. Here culture grades into science, since most of the Paleolithic archaeologists were also geologists. In biology itself, Wallace and others would promote the biogeographical evidence for a theory of common descent and for the transforming effects of migration, isolation, and interspecific competition. Morphologists would have become increasingly aware of the evidence from comparative anatomy and embryology suggesting evolutionary relationships between groups. Key fossil discoveries encouraged a growing sense that the history of life was best understood in terms of trends rather than isolated creative acts.

The basic structure of the evolutionism that would have emerged is surprisingly easy to reconstruct because many of its components played important roles in the development of our own science and culture. They include some concepts that were also exploited by Darwin, but also others that became popular even though they did not sit very well with the theory he articulated. Darwin's theory did not dominate the evolutionism of the late nineteenth century, and even those "Darwinian" components that did gain credence were often only peripherally related to what modern science regards as his core theory of natural selection. So it is much easier than one might expect to extract the selection theory from the mix and work out what a non-Darwinian evolutionism would look like. The theory of common descent would be important, though heavily qualified by efforts to visualize evolution as the ascent of a linear scale of development leading to European races and Western culture. Darwin accepted that evolution was progressive in the long run, but he knew that progressive steps were rare, and he appreciated that many different branches of the tree of life had advanced to higher levels of organization. This point was all too often ignored by those who wanted to see themselves as being at the highest point of development, with all lesser races and species being demoted to side branches on the tree of life. Progress all too often trumped local adaptation as an explanatory tool, and many biologists refused to accept that the key steps in evolution could be explained as a series of local adaptations. Biogeography was important, however, because it legitimized the idea that the less advanced species and races tended to be eliminated whenever a higher type invaded their territory. Thus the idea of a struggle for existence at the racial and species level would emerge even in a world lacking the theory of natural selection acting at the individual level.

CULTURAL EVOLUTION

The transformation of religious and political views continued into the 1860s. Some of the cultural changes popularly attributed to Darwin's influence could have been linked to any version of evolutionism and are in fact products of values imposed on science, rather than derived from it. As James Secord notes in the conclusion to his study of the impact of Chambers's *Vestiges*, from the perspective of popular culture, Darwin merely completed the revolution begun by the earlier evolutionary text. It follows that the revolution would have run its course anyway, although perhaps Darwin's intervention sped things up by forcing scientists to climb off the fence and throw their weight behind the evolutionary bandwagon. Without Darwin, what Secord calls the "popular science of progress" would still have become the dominant ideology under the influence of Spencer and other radical thinkers, and sooner or later the scientists would have had to take note.

The Liberalization of Belief

A key feature of the new ideology was the increasing liberalization of religious belief associated with a decline in the acceptance of biblical literalism and the doctrine of miraculous creation. Few could now be unaware of the vast sequence of earth history revealed by geology and palaeontology, and it was difficult to imagine that the vast numbers of new species that had appeared throughout history had been formed by direct acts of the Creator. It was also obvious that there were trends in the sequence of fossil forms within each family, so even if the individual species were created by miracle, there was a definite pattern linking them into a divine plan. For many deeply religious scientists, the law-like nature of creation would become a central feature of an evolutionary philosophy. The anatomist Richard Owen, widely but incorrectly dismissed nowadays as an opponent of evolutionism, emerged as a champion of what is called theistic evolutionism—the idea that evolution is the unfolding of a divine plan built into the laws instituted by the Creator. Charles Lyell and Asa Gray, both supporters of Darwin in our world, were influenced by this synthesis of evolutionism and the argument from design. For others, theistic evolutionism formed a bridge leading toward a Lamarckian theory in which individual effort and initiative directed evolutionary change. Although a purely natural process, the inheritance of acquired characteristics seemed to be just the sort of mechanism a wise and benevolent God might use to achieve His goal via progressive evolution.

Liberal clergymen too were increasingly willing to seek such a compromise with the new discoveries of science. A good example is Charles Kingsley, in our world a clerical supporter of Darwinism. His popular book *The Water Babies* of 1862 lampooned the scientific debates but had the serious purpose of trying to convince his readers that God has given us the power to improve ourselves—and thus to contribute to the progress of the human race. Because Kingsley praised Darwin, his work is often seen as Darwinian, but its philosophy of progress through effort was purely Lamarckian. James Moore has shown that many liberal clergymen in Britain and America became followers of the Spencerian philosophy, in effect following Kingsley by transforming Lamarckian evolutionism into a form of muscular Christianity.[5] These were not easy transformations, because conservative religious thinkers still found it hard to accept the demise of traditional Christian beliefs and values. The controversy sparked by the 1860 collection of liberal theological essays entitled *Essays and Reviews* was even more heated than that over the *Origin of Species*. But the very fact that Darwin's theory was not the only influence promoting the liberal cause within the churches suggests that the move to reject miracles in favor of divinely instituted laws of creation had a momentum sourced from outside science.

Liberal Ideology

It seems odd that clergymen could ally themselves with Spencer's philosophy, often seen as a secularizing influence on nineteenth-century thought.

Spencer and Huxley were allies in a campaign to eliminate miracles and design as components of educated thinking about science. Evolutionary naturalism, as it became known, was offered as an alternative to formal religion. It provided a naturalistic source of ethical values based on a supposedly impartial study of nature that presumed that all changes—physical, biological, and social—are governed by law. For Huxley, this approach also helped to promote the cause of professional science as a source of expertise in a modern society. Naturalism was intended to replace organized religion and was often presented in almost evangelical terms. But Huxley and Spencer were anxious to avoid the charge that they were atheists. Huxley coined the term "agnosticism" to denote his position, while Spencer's philosophy included an "Unknowable" lying behind the world we actually observe. The gap between the new secularism and liberal religion was not as great as it might seem, and both sides could seize on the idea of evolution to explain how the world had developed without direct supernatural intervention.

Spencer's writings did not sell well in the 1850s, but the publication of his *First Principles of a New Philosophy* in 1862 inaugurated his "system of synthetic philosophy," which would soon become immensely influential. *First Principles* seems very dry reading today, but for Huxley and many other forward-looking thinkers, it seemed to codify a secular way of looking at the moral and physical worlds. The key was the idea of evolution, seen as an inevitable process of development driven by natural law. For Spencer this implied not only an increase in the complexity of individual systems, but also a multiplication or diversification of the number of systems. In principle his vision of evolution did not imply progress toward a single predefined goal. In practice, however, the vision was easily subverted into a developmentalism that saw evolution as the ascent of a linear hierarchy, all deviations from the main line being mere side branches of no ultimate consequence. Spencer thought that the same laws drove all levels of evolution, so the transition from animal evolution to human social evolution was seamless and involved no new mechanisms. Although life had evolved a multiplicity of complex forms, only one—the human—had allowed the transition to the social level of progress. Social progress itself was also presented as an inevitable trend that in recent times produced the transition from feudalism to free-enterprise capitalism.

The system of synthetic philosophy was expounded in a series of books of which *Principles of Biology* was the first to appear in 1864. Huxley had interacted with Spencer in the writing of this book, and it is hard to believe that the process would not have forced him to rethink his attitude toward biological evolution. Admittedly, Huxley had no time for the Lamarckian principle at the heart of Spencer's biological and social evolutionism, but there were other mechanisms to which a biologist concerned with morphological differentiation rather than local adaptation could turn. For Spencer, it was the inherited effects of the individual's purposeful efforts to deal with environmental challenge that drove both biological and social evolution. In our world, *Principles of Biology* added references to natural selection in which Spencer coined the iconic phrase "survival of the fittest." But the book is still more Lamarckian than Darwinian, as befits the prelude to a philosophy of social evolution in which competition drives individual self-improvement. For Spencer, "fitness" seems to have included adaptability, so natural selection could only follow up the Lamarckian effect—Spencer did not share Darwin's view that much variation is undirected.[6]

In a world without Darwin, Spencer's Lamarckism would have encouraged some younger naturalists to begin the process of trying to turn the general idea of progressive evolution into a workable scientific theory. He certainly prompted a whole generation to enthuse over the promise of social evolution driven by individual effort, enshrining the positive effect of competition as the underlying motor of change. As Michael Ruse has suggested, few were converted to evolutionism by the scientific evidence alone—it was enthusiasm for the idea of progress that encouraged many to extend the idea back from human society into the animal kingdom.[7]

Parallel developments were taking place in Germany, although here there was less interest in the principle of laissez-faire at the heart of Spencer's political system. The old German tradition of transcendental philosophy had encouraged the view that all living forms were part of a coherent pattern. This vision was consistent with a kind of formalized evolutionism, although it seldom openly promoted transmutation in practice. The followers of J. F. Blumenbach were quite willing to search for alternatives to divine creation, and in the middle decades of the century their attention was beginning to switch from autogenesis (spontaneous generation) to transmutation. In the middle of the century a group of materialist thinkers challenged traditional values even more openly than the British exponents of scientific naturalism. Biologists such as Karl Vogt, anxious to promote alternatives to the idea of divine creation, were now beginning to pin their hopes on evolutionism. For young, radical scientists in the 1860s, including Ernst Haeckel, the idea of progressive evolution driven by natural law was the obvious basis for a synthesis of biology and social thought, paralleling the somewhat different version of this way of thinking that Spencer offered to the English-speaking world. Here too Lamarckism would be a prominent component of the effort to explain how species differentiated, just as cultures pass on improvements to the next generation.

Spencer's *Principles of Psychology* had already presented an evolutionary view of how the faculties of the human mind had been acquired through evolution, and Haeckel too was keenly interested in human origins. In effect, Spencer's approach to psychology bridged the gap between the biological and social phases, although his book took such an abstract approach that it did not reopen the debate—simmering since even before *Vestiges*—over the link between humans and an animal ancestry. Two events in the period around 1860 did prompt a reassessment of human origins and would have done so whether or not there had been an initiative from biology. The growing enthusiasm for a model of human history based on the idea of progress was boosted by a revolution in archaeology that opened up the prehistoric past and emphasized the primitive origins of technology, culture, and society. Historians used to depict this revolution as a spin-off from the one sparked by the *Origin of Species*, but it is now evident that it took place quite independently of Darwin's initiative. At the same time, the debate over humanity's relationship to the great apes was reinvigorated by fresh anatomical studies.

Prehistoric Archaeology

Despite the extension of the geological timescale, it had remained an article of faith that no human remains were more than a few thousand years old, leaving the biblical model of human origins unchallenged. Discoveries of primitive stone tools alongside the remains of extinct animals were dismissed as fraudulent. In the late 1850s a group of British geologists reexamined the sites where these discoveries had been made and became convinced that they were genuine. Suddenly the bottom dropped out of human history. The ancient civilizations recorded in the Bible had been preceded by a "prehistoric" period of immense duration in which our ancestors had lived as primitive savages. The implications of these revelations,

102 · CHAPTER FOUR

summarized for a general audience in Charles Lyell's *Antiquity of Man* in 1863, were obvious. There might be no "missing link" between humans and apes in the fossil record, but it was now much easier to imagine that the earliest humans had emerged from an ape ancestry. The archaeologist John Lubbock, who coined the terms "New Stone Age" and "Old Stone Age" to denote the progressive refinements in tool making, openly linked the progress of human culture into an evolutionary perspective for the history of life. This new perspective slotted neatly into Spencer's model for the gradual emergence of human mental faculties as a Lamarckian response to environmental challenges.[8]

In our world Lubbock was a supporter of Darwin, but his vision of a ladder of cultural evolution leading toward the present "advanced" state of society was not modeled on the idea of common descent. Lubbock also contributed, along with Edward B. Tylor and Lewis Henry Morgan, to the emergence of a new evolutionary anthropology. Here studies of "savage" societies such as the Australian aborigines presented them as primitive relics of the oldest Stone Age cultures, in effect allowing anthropology to flesh out the lives of the makers of the ancient stone tools.[9] The savage was a relic of the past, a living fossil isolated from the mainstream of progress and thus throwing light on how our earliest ancestors had lived. This view was not derived from Darwinism—Lubbock and Morgan endorsed biological evolution, but Tylor did not. Instead, he proclaimed the unity of human nature around the globe, in opposition to an ever-growing tendency for physical anthropologists to depict the non-European races as mentally and biologically inferior. Far from being a product of Darwinism, this model of cultural evolution promoted a linear, goal-directed vision of progress profoundly at variance with Darwin's "branching-tree" model.

There were no fossils of the ancient toolmakers available in the 1860s, but physical anthropologists were increasingly willing to depict the "lowest" human races as mentally and biologically inferior to Europeans. They used measurements of their cranial capacity—now widely dismissed as the product of either deliberate or unconscious manipulation of the evidence—to argue that they had smaller brains and hence by implication lower levels of intelligence. Physical anthropologists such as Robert Knox who rejected biological evolution nevertheless suggested that the lower races also had features that were distinctly apelike. This interpretation of racial differences preceded the emergence of evolutionism, although anthropologists who favored a linear model of progress took it up with enthusiasm. In the absence of fossils, the savage races of today became living exemplars of the "missing link" between modern humans and our ape ancestors. The debate over the relationship between human and apes had raged in earlier decades, sparked by the evolutionary theories of Lamarck and Chambers. Now it was reignited—but was this solely owing to Darwin's new proposals on biological evolution? The revolution that took place in ideas about human prehistory suggests that it was not. In a world without Darwin, the question of human origins would have come to the fore in the 1860s anyway and would have forced biologists to reconsider the question of evolution.

Apes and Humans

Another high-profile issue focused attention even more closely on the relationship between humans and apes. Since the late 1840s Richard Owen had been working on the anatomy of apes, insisting that there were major differences between them and humans. His comments made it clear that this gulf was a challenge to the evolutionary explanation of human origins offered by Lamarck and Chambers. Like many scientists with religious scruples, including Lyell and Wallace, Owen was tempted by the general idea of evolution but wary of applying it to humans. In the 1850s, the skeletons and skins of gorillas at last began to reach Europe, rekindling interest in the relationship. The traveler Paul Du Chaillu supplied some of these specimens and produced a bestselling book on the topic in 1862. In order to boost his collections at the British Museum, Owen exploited contacts with Du Chaillu, a relationship that backfired to some extent when the latter's accounts of the ferocious behavior of gorillas were challenged by later explorers. Owen continued to publish on the topic, insisting that apes and humans were so distinct anatomically that they should not be included in the same order, the Primates. He separated the human species into a distinct order that he called the Bimana (two-handed) as opposed to the Quarumana (the apes, which in effect had four hands because their foot structure was not as distinct as in humans). He also coined the term "Archencephala" for humanity, emphasizing the dominant power of the brain. For Owen, the enhanced structure of the human brain was a divine gift responsible for our higher intellectual and moral powers.[10]

Enter Huxley, Owen's great rival, determined to challenge his intellectual and institutional authority.[11] In 1858 Huxley began an attack on Owen's descriptions of the cerebral anatomy of apes, pointing out what he presented as serious inaccuracies. He challenged the claim that a particular structure, the hippocampus minor, was absent from the ape brain. In our world, this debate was soon linked with that over Darwinism, although Darwin had hardly mentioned human ancestry in the *Origin of Species*. Huxley clashed with Owen on ape brains at the same meeting of the British Association in 1860 in which (in another session) he had his classic encounter with Bishop Samuel Wilberforce over Darwin's theory.

Huxley's *Man's Place in Nature* of 1863 is widely supposed to have confirmed his triumph over Owen on ape anatomy, and to have boosted the Darwinian cause. Yet Huxley made few explicit references to human evolution; his book argued merely for the close anatomical resemblance between humans and apes and for the older view that they should be included in a single order, the Primates. The implications were obvious to all, of course. But they would have been obvious even if Darwin had not published because Owen had made no secret of why he opted for human uniqueness. And Huxley would almost certainly have challenged him on the topic of ape anatomy whether or not there was a new debate over the general idea of evolution. Coming as it did around 1860, this debate provides another good reason for thinking that Huxley and others would have been led to reconsider their position on evolution in a world without Darwin. Proclaiming the close relationship between humans and apes would have identified Huxley with an evolutionary position in the public mind, whether he intended this or not.

Here a survey of changing attitudes has brought us right to the heart of science, confirming the interpenetration of scientific and cultural developments. Owen seemed happy to pin the distinction between humans and apes on anatomical differences, in effect accepting the materialist position that our higher mental faculties were the product of a bigger and more complex brain. The traditional view that humans alone possessed an immortal soul, a divinely implanted spiritual faculty that distinguished us from the "brutes that perish," was no longer secure. Whether one looked to cerebral anatomy, psychology, social and cultural development, or archaeology, perspectives favorable to an evolutionary account of human origins were gaining strength around 1860. Coupled with the growing distrust of miracles and the popular enthusiasm for the idea of progress though law, it can plausibly be argued that everyone's attention, including that of scientists, would become focused on the topic of evolution in the 1860s, even without an input from Darwin.

THE TRANSFORMATION OF SCIENCE

How do we work out how the scientific world of the 1860s would have responded to a less dramatic introduction of the idea of evolution? Two sources of evidence that can throw light on the issue are surveys of changing attitudes to Darwinism in our own world and biographical analysis of the work done by key figures who might have contributed to the change in thinking, whether or not inspired by the *Origin of Species*. There are a huge number of historical surveys of the Darwinian revolution, allowing us to establish a number of generalizations.[12] These insights allow us to see that in a world without Darwin there would be other initiatives to explore the scientific opportunities that the theory of common descent offered.

Too wide a generalization is not possible, however, and the reaction to Darwin was by no means uniform around the world or even within the different branches of the scientific community. The professionalization of that community is also a factor. Huxley and the new generation of dedicated "men of science" seized on Darwinism as a weapon to use against those who still wanted science to remain subordinate to the church. Without Darwin the radicals would still have been able to use the theory of common descent as ammunition, but without natural selection their campaign would have aroused less hostility among the defenders of the argument from design. There were also many less radical thinkers anxious to modernize traditional values by taking the basic idea of evolution on board.

How many scientists would it have taken to swing the balance? We know that Darwin was extremely concerned in the early years of the debate—he hoped to convert a few key figures in order to gain a foothold while the broader community had time to come to terms with his new theory.[13] There was much initial hostility in some quarters, but a broader conversion occurred by the end of the 1860s. Articles and books supporting the general idea of evolution began to outnumber those opposing, although only a very few expressed unqualified support for natural selection. The same pattern is observable in Britain, America, and Germany, although the details of who supported evolutionism and why differ.

Ernst Haeckel later insisted that there were only a handful of German supporters when he began his campaign for Darwinism in 1863—yet he had written to Darwin in July 1864 to say that the younger generation of naturalists was increasingly supportive.[14] The discrepancy probably reflects the fact that while there was little enthusiasm for divine creation, naturalists were only slowly coming to regard evolution as the most plausible alternative. Recent surveys identify a significant number of positive reactions to Darwin, although again they include many who did not accept natural selection.[15] There were serious barriers to a full understanding of Darwin's theory in Germany, most obviously a lack of familiarity with the evidence he presented from animal breeding. The rise of non-Darwinian evolution theories over the next few decades makes it clear that the selection theory, far from being the reason why people accepted evolution, was actually a hindrance for many.

Countries and Communities

Evolutionism meant different things in different countries. Its reception also depended on the scientific specializations of those involved, not necessarily in ways one might expect. More seriously, there were radical, liberal, and conservative factions within the scientific disciplines, just as there were in the wider community. We assume that it was the radicals and liberals who jumped most easily onto the evolutionary bandwagon, but this is not necessarily the case. The more imaginative conservatives could find ways to update the teleological worldview by seeing evolution as an expression of the Creator's purposes. These ideological divisions cut across scientific specializations, and in some cases it was personal and professional loyalty that dictated how a particular individual would react.

Haeckel's 1864 letter supports the widely held view that the scientists who signed up for Darwinism tended to be from the younger generation. This assumption has its critics,[16] but there does seem to be something in it. Older figures committed to the idea of divine creation, including Adam Sedgwick and Louis Agassiz, were never going to give ground. Younger workers in some areas of science were equally inflexible. Experts specializing in the classification of particular groups of animals and plants were often unsympathetic to a theory that challenged their neat pigeonholing of species. Physicists were hostile because the theory didn't square with their rigorous standards of scientific explanation. But if we survey the whole swathe of relevant sciences, the intuition of Darwin and Haeckel that it was the younger members of the communities who displayed the greater flexibility in the face of the challenge seems to hold up. Several geologists from the newly established Geological Survey of Great Britain came out in support. They were followers of Lyell, and to them Darwin's methodology of using evidence from the present to reconstruct past events made perfect sense.

William Montgomery makes the interesting point that in Germany it was the invertebrate zoologists who were most likely to become Darwinians, noting that these were members of the younger generation trying to make their name in what was a relatively new discipline.[17] Human and vertebrate anatomists in medical schools were better established and often more hostile. Some of the younger zoologists supported Haeckel in his efforts to use evolutionism to attack traditional religious values. But there were also some biologists, including major figures such as Rudolf Virchow, who saw evolution as the expression of a purposeful progressive force. Haeckel tried hard to get evolutionism identified with liberalism and materialism, but there were many conservative thinkers flexible enough to realize that the idea of supernatural design could be expressed in the forces driving evolution. In a world lacking the selection theory to highlight the materialistic interpretation of evolutionism, the division between these two camps might have been much less acute, allowing a more unified evolutionary program to emerge.

The same situation held true in Britain, although historians have lost sight of the conservative program. On the radical side, Huxley's supporters included several younger vertebrate anatomists who gained appointments at just the right time, including George Rolleston at Oxford and W. H. Flower in London. Here, as in Germany, morphology—the study of organic form, both vertebrate and invertebrate—became a key focus of evolutionary study. The idea of common descent explained the relationships between structures revealed by anatomy and embryology. It did not necessarily encourage the view that the modifications were due to adaptive pressures from the environment, though, and even Huxley remained suspicious of Darwin's emphasis on adaptation. The evidence from biogeography was of less interest to this community.

Huxley was anxious to exploit evolutionism as a means of discrediting outdated naturalists still wedded to natural theology, but without the selection theory this aspect of his campaign may have been less significant. We tend to forget that his efforts to liberalize and professionalize the scientific community were not universally successful. There were anatomists such as Richard Owen who had been trying for some time to update the argument from design by suggesting that the Creator's purposeful intentions were implanted in the laws governing the development of life on earth. Owen and his followers were able to establish a more conservative evolutionary program, and—as in Germany—in a world lacking the selection theory, the gulf between the liberal and conservative camps would have been much less divisive.

Some palaeontologists were hostile at first, only too well aware of the discontinuities in the fossil record. But even relatively conservative figures such as Owen and H. G. Bronn were aware of systematic trends in the sequences of fossils in some groups. By the early 1870s new discoveries were lending support to Darwin's argument that the gaps in the record were due to lack of evidence. The input from these discoveries should not be exaggerated—the reptile-bird intermediate *Archaeopteryx*, described by Owen in 1863, was not at first seen as evidence for evolution even by Huxley. But by the end of the decade, Haeckel persuaded Huxley to begin tracing phylogenies (lines of evolutionary descent) in the fossil record. Several of Agassiz's young disciples switched to evolutionism in the late 1860s on the strength of their palaeontological studies. But theirs was a distinctly non-Darwinian form of evolutionism, allowing Alpheus Hyatt and Edward Drinker Cope to go on to become founders of what became known as the American school of neo-Lamarckism.

In the English-speaking world, biogeography provided the most convincing evidence, perhaps because it was more visible in a nation committed to worldwide trade and exploration. Darwin included two chapters on the topic in the *Origin of Species*, but others were looking in the same direction. Important figures such as Lyell noticed Wallace's 1855 paper implying common descent, and Wallace continued to add to the biogeographical evidence in a series of papers and a massive 1876 book, *The Geographical Distribution of Animals.* In botany, Joseph Hooker's and Asa Gray's studies of the global distribution of flora brought them ever closer to the idea of divergent evolution caused by geographical dispersal. For both, it was common descent that offered the real benefits to the science of dispersal, not the details of how Darwin imagined transmutation to work.

Darwin had coached Hooker, of course, and Gray heard about natural selection in 1857, shortly before Wallace penned his paper triggering publication of the theory. But when we survey their efforts to make sense of the problems posed by the distribution of species, we see that there are good reasons for supposing that even without Darwin's prompting, these naturalists would have been encouraged to explore the idea of common descent over the course of the 1860s. This in turn would have prompted morphologists such as Huxley and Haeckel to think about whether the same theory could help them to understand structural resemblances and the trends in the fossil record. Innovative figures in these two areas would thus have pushed the scientific community as a whole to reconsider its position on evolution in the course of the 1860s. Without Darwin, there would have been no major shock to the system and no theory of individual natural selection. But there would have been a steady drip of suggestions pointing to the general idea of common descent, eventually wearing away the resistance of all but the most conservative naturalists.

Surveying the known work of key figures allows us to reconstruct how they might have been led toward an evolutionary perspective in a world without Darwin. Cultural and scientific pressures were forcing them in this direction, but they could gain most of what they needed without the idea of natural selection. In fact, most of them found selection a hindrance rather than a help in their efforts to grapple with the scientific issues. For some it was also a hindrance in terms of their religious and moral beliefs. Radical figures such as Huxley and Haeckel, who welcomed natural selection as a weapon in their campaign against the influence of organized religion, would have had one less string to their bow. Yet even for them, a naturalistic model of common descent would serve most of their purposes in challenging the conservative view of creation. Without Darwin's highly materialistic hypothesis, they would have explored other mechanisms that would have been less offensive to their opponents. In this situation it would have been easier for a conservative such as Owen to present himself as a supporter of evolutionism, heading off the public relations disaster that has led to his unfair dismissal as an opponent to this day.

The Biogeographers

The traveling naturalists who studied the geographical distribution of species around the globe were, in effect, following in Darwin's footsteps. Wallace is the most obvious example, not because of his somewhat cryptic 1858 paper on natural selection but because his 1855 paper clearly pointed the way to the theory of common descent and alerted a key figure, Charles Lyell, to that possibility independently of any input from Darwin. A survey of Wallace's activities from this date into the 1870s suggests how he might have played his hand in a world without Darwin. Hooker and Gray merit similar attention, because we know that their work was generating puzzles that could be solved by the theory of common descent and because—unlike Wallace—they were influential figures at the heart of their respective scientific communities.

Imagining a counterfactual Wallace poses difficulties both because of the debates about the content of his 1858 paper and because he was always something of an outsider to the elite scientific community.[18] He was a collector and minor author who never gained a significant professional position and was viewed with increasing suspicion when he took up unfashionable enthusiasms such as spiritualism, land reform, and the anti-vaccination campaign. As he himself admitted, he was never going to serve as the leader of a power block able to wrest control of the scientific elite from the older generation. But this does not mean that he would have been unable to exert an influence from the margins, given that his new ideas on biogeography and the relationship between species and varieties were responding to issues puzzling many among the elite.

Given the ambiguities in Wallace's 1858 paper, it is unlikely that his subsequent work would have focused on selection at the level of individual variation. He certainly would not have included an extensive study of animal breeding and the clues that could be obtained from it. Wallace's early papers focus on the difference between varieties and species (or, rather, on the lack of real difference) and the tendency of a single form to divide into a number of descendants, especially when populations become separated by geographical barriers. His 1855 paper had outlined, in effect, the theory of common descent, although he made no effort to clarify how the divergence took place. Although initial response was muted, the paper did begin to make an impression on influential figures such as Lyell, whose theorizing in geology had served as a model for Wallace's thinking (as it did for Darwin). The 1858 paper invoked selection when conditions become harsh and population pressure leads to the extinction of those varieties and species less adapted to the prevailing environment. Given the lack of response to the joint Darwin-Wallace papers in our world, it seems doubtful that this short piece would have had much impact even if it had achieved publication.

What would have had an effect is the series of more specialized papers on geographical distribution and what we now call speciation that Wallace published both before and after his return to Britain in 1862. In the late 1850s he had thought about preparing a book on the topic, and perhaps in the absence of the *Origin of Species* he would have gone ahead with this project. He gained considerable credit with the scientific community for his definition and explanation of the line that bears his name separating the Asian from the Australian faunas on the islands of what is now Indonesia. Later on, as more information became available from around the globe, he would publish major studies of biogeography, his *Geographical Distribution of Animals* in 1876 and *Island Life* in 1880. These were hugely important surveys triggering an explosion of interest in the topic in the later decades of the century.

How effective would Wallace's papers (and hypothetical book) have been in promoting the idea of common descent in the 1860s? In the previous decade he had been an enthusiastic follower of Spencer's philosophy and a determined advocate of naturalism; his challenge to the stability of species was intended to undermine the implication that some form of supernatural intervention was needed to create them. By the 1860s he began to adopt an approach that was much more sympathetic to the traditional view that the history of life unfolds according to a divine plan. This came out most openly when he explicitly began to argue that the human mind could not have evolved by purely natural means. In our world, this move distanced him from Darwin and his followers, but given Wallace's conversion to spiritualism it seems clear that he would have taken this position anyway. His last major book on evolution, *The World of Life* (1910), presented the whole history of life on earth as divinely preordained. These developments in Wallace's worldview began in the 1860s and must be taken into account when trying to imagine how he would have presented his technical papers on natural history at that time.

Despite its emphasis on the struggle for existence, Wallace's 1858 paper does not clearly state that the process that divides a species into varieties is driven by the adaptation of those varieties to local conditions. Varieties go extinct when harsh conditions put pressure on those that are less well adapted to the environment as a whole. As Martin Fichman has pointed out, there are sentences in Wallace's early papers implying that at least some of the characteristics that distinguish varieties are non-adaptive and may have their origin in supernatural design.[19] His classic 1864 paper on the Malayan Papilonidae (a group of butterflies) notes a local influence on the formation of varieties but implies that its effects are "unintelligible" and "mysterious."[20] These are not words that would be used by someone who was convinced that the whole process was driven by adaptation and natural selection acting at the individual level.

Wallace certainly realized that some of the butterflies' colors were adaptive, and the one feature that might have driven him to more detailed study of individual natural selection is mimicry. Henry Walter Bates, his former traveling companion in South America, pioneered the study of mimicry, by which an edible species acquires the same warning coloration that an inedible species uses to warn off predators. In 1862 Bates pointed out that such effects are best explained by natural selection.[21] Since insects do not have control over the color of their wings, Lamarckism via change of habit is out of the question—although the Lamarckians also postulated intrinsic adaptive responses by the organisms' physiology to changed conditions. Wallace too picked up on mimicry and made other studies of animal coloration (he disagreed with Darwin's belief that some colors played a role in sexual selection).

Would Bates have come up with the explanation of mimicry in terms of natural selection if there had been no *Origin of Species* for him to read? This seems unlikely, and if he did not, would Wallace have taken up the subject? In a world without Darwin, he may have concentrated on dispersal and divergence, leaving the actual formation of varieties as a problem for future research. Under the influence of Spencer, he might even have allowed a role for the Lamarckian effect, something he rejected in our world once Darwin alerted him to the action of natural selection on individuals.

For the non-Darwinian Wallace, selection would have been invoked only to explain the disappearance of poorly adapted varieties that had been formed by as yet unknown processes. This would allow him to go on to write *The Geographical Distribution of Animals*, in which he saw hardier species evolved in northern latitudes continually moving south and displacing the less advanced species inhabiting those regions.[22] His was not a worldview without competition, but it would lack Darwin's emphasis on the relentless and brutal effects of individual struggle.

Wallace would play a significant role in alerting other naturalists to the power of the theory of common descent as a means of demystifying the nature of species and explaining both the relationships used to classify species and the dispersal of species around the globe. He would not have promoted natural selection, except as a negative force eliminating the less well-adapted species and varieties. He certainly would not have served as a figurehead for a movement to transform the scientific community—he would be a gadfly, not a leader. When he returned to Britain, he attended meetings of the British Association and met leading figures such as Lyell and Spencer. But his favorite haunts were the societies where travelers and amateurs were still active, including the Royal Geographical, the Entomological, and the Zoological Societies. When he began to develop his ideas on the antiquity of the human races, he presented them at the Anthropological Society of London, a hotbed of the most extreme form of race science where figures such as James Hunt proclaimed the races to be distinctly created species. Darwin, Hooker, and Huxley would have nothing to do with this extremist group. Even before Wallace turned to spiritualism and postulated supernatural intervention in human origins, he had alienated the elite group with which Darwin was so anxious to work. He remained an outsider for the rest of his life, simultaneously respected for his contributions in some areas of science but marginalized for his eccentricity.

The people Wallace most needed to convince were those who had interests in the topic of species and their distributions, most importantly Lyell and Hooker. Both were leading figures in the scientific community, but they were of different generations. Lyell's methodology requiring the explanation of past geological changes in terms of observable causes inspired Darwin, Wallace, Hooker, and other naturalists to investigate how anomalies in the distribution of animals and plants could be explained in similar terms. Hooker, like Darwin, extended this position into a complete scientific naturalism. He was a member of the X Club, an unofficial group of liberal thinkers including Huxley, Lubbock, and Spencer who wanted to eliminate the role of religion from the discourse of science. This group provided a forum through which professionals like Huxley could lobby to get their younger disciples into influential positions.[23] Lyell was not a member, because he retained the values of an earlier age and found it difficult to extend his methodology into the realm of biology, where it might challenge the idea of divine providence and the uniqueness of humanity's spiritual status.[24]

Wallace moved from an earlier naturalism toward a position much more in tune with Lyell's. In the non-Darwinian world, Lyell would have found it comparatively easy to come to grips with Wallace's approach to evolutionism, because it did not entail the radical challenge to traditional values that Darwin's theory posed. We know that Lyell was impressed by Wallace's 1855 paper—indeed, it is the first work he referred to when he opened his notebooks on the species question in 1855.[25] But as he subsequently went on to grapple with Darwin's theory, he found it difficult to accommodate its complete rejection of design. Lyell wanted a natural theory of species production, but he also wanted to believe it was somehow purposeful. One could preserve the necessary level of evasion if one postulated an unspecified "variety and race-making force," which is exactly what Lyell might have seen (in the absence of any input from Darwin) in Wallace's 1858 paper.[26]

For thinkers such as Lyell, torn between the old supernaturalism and the strident new radicalism, it was useful to avoid considering exactly how new forms originated. The benefit of evolutionism came from the implications of common descent for classification and of the principles of divergence and dispersal for biogeography. Worrying about how a localized population became adapted to its environment might raise all sorts of awkward questions. Lyell, Owen, and many others who faced this dilemma would have found the introduction of evolutionism far more congenial if they were not forced to grapple with a radically naturalistic theory of transmutation. Wallace's approach might have appealed to them for just that reason.

Hooker, by contrast, was a key ally of Huxley's in the fight to promote scientific naturalism and win a better position for professional "men of science." He had no time for those who would retain elements of the old teleology and was as critical as Darwin when Wallace turned to spiritualism and the supernatural. He was ideologically primed to welcome a naturalistic theory of evolution such as Darwin's, just as his scientific background prepared him to see the logic of the broader argument offered by Darwin and Wallace. Unlike Lyell, he had no theological qualms about the theory of natural selection, but as far as his scientific work was concerned, the basic idea of divergence from a common ancestor would suffice to deal with most of his problems. Hooker's role in a world without Darwin is difficult to assess precisely because in the real world he was so close to Darwin. He was the first naturalist to be informed about the theory of natural selection, and by the time he came to write the *Origin*, Darwin had been prodding him for over a decade to overcome his initial doubts about the theory. How far would he have come toward evolution by the 1860s if Darwin had not prompted him?

Whatever his reservations about Wallace's broader views, Hooker respected his work on biogeography. The feeling was mutual, with Wallace dedicating his *Island Life* to Hooker. But this interaction only began to flourish in the 1860s and 1870s, and there is no evidence that Wallace's 1855 or 1858 papers offered Hooker—who had, unlike Lyell, been privy to Darwin's theory since 1855—any important insights. In a world without Darwin, however, Lyell's enthusiasm for Wallace's 1855 paper might well have encouraged Hooker to press on with his own investigations of how best to understand the relationships between species and their distribution in space. His own explorations in India and across the Southern Hemisphere had left him with a wealth of information on the distribution of plants that (in our world) Darwin was able to use to convert him to evolutionism. Without Darwin, and with possible encouragement from Lyell, Wallace, and Spencer, Hooker might well have moved in the same direction, if more slowly.

Historian Jim Endersby argues that there was another reason why Hooker came to see the value of the theory of common descent: it helped him to deal with the practical problem of controlling the relationship between species and varieties.[27] From the Royal Botanical Gardens at Kew, Hooker administered a global network of collectors, many of them amateur enthusiasts who regarded every slightly different form discovered in their own region as a distinct species. To make classification manageable, Hooker often wanted to lump a series of such forms together as local varieties of a single species, and Darwin's theory gave him a rationale for doing this. If varieties were incipient species, it was easy to show how an enthusiast could exaggerate the degree of difference and imagine that true speciation had already taken place. Yet Hooker needed to be able to treat species as fixed units for the purposes of classification, so he was only able to accept evolution on the understanding that it occurred so slowly that it made no practical difference to this area of science. Natural selection thus appealed to him because Darwin insisted that it worked slowly—but any other mechanism conceived to act at the same rate would have similarly appealed to him. A theory in which episodes of rapid change were interspersed with long periods of stability would also have filled the bill.

Although Hooker became a Darwinian, his real enthusiasm was for the idea of divergent evolution. He explicitly complained to Darwin that he had exaggerated the role of natural selection.[28] In a world without Darwin, Hooker might have combined his own insights with those of Wallace to focus on a less clearly defined process of divergence, probably bringing in Wallace's notion of varieties being driven to extinction at times of environmental stress. Spencer might have encouraged him to believe that there were ways in which the physiology of plants could react to environmental changes, but as a taxonomist Hooker would have had little incentive to explore such processes in detail. Coupled with his own work suggesting the importance of dispersal in bringing invasive species into new localities, the idea of divergent evolution would have given him most of what he needed to push his science forward. As Endersby notes, Hooker's science did not involve any dramatic transformation of technique—evolution offered a new justification for what he had been doing anyway. In the counterfactual world, we can imagine Hooker playing a significant role in the gradual conversion of the scientific community to evolutionism during the course of the 1860s without invoking the idea of natural selection acting at the individual level.

Like many botanists, Hooker was impressed by Darwin's explanation of how the structure of plants, especially their flowers, could be seen as products of an adaptive process. Richard Bellon has argued that this aspect of Darwin's work provided key support for his general campaign, impressing even those botanists who had been reluctant to think in evolutionary terms.[29] In the counterfactual world, this evidence would not have been explored at the time, possibly slowing down the general conversion to evolutionism. Its absence would also remove a key element in the case for adaptation being the driving force of change. But even without this line of argument, I suspect that biogeography would have driven those botanists concerned with classification and distribution toward the general idea of divergent evolution.

In America it was another plant geographer, Asa Gray of Harvard, who—along with geologist William Barton Rogers—provided the initial support for Darwinism. Gray had known about Darwin's theory of natural selection since 1857 and had taken it on board to bolster his campaign against the idealist worldview promoted by Louis Agassiz. As an extreme creationist, Agassiz presented a far more obdurate opponent to evolutionism than Richard Owen did in Britain, and we can only understand the American situation in the context of the opposition by more empirically minded scientists to his influence. Gray stood opposed to Agassiz on this front and on the vexed question of the relationship between the races of mankind, then driving the United States toward war. Yet, unlike Hooker, he had no time for the naturalistic philosophy, holding to a staunch Presbyterianism and resenting the influence of Spencer (whose a priori approach to knowledge he saw as little better than idealism). Gray's passion was for the evidence, and his studies of the distribution of North American and Japanese plants led him to become increasingly dubious about the fixity of species, increasingly sure that their distribution could only be explained by migration rather than creation. Gray strongly resisted Agassiz's claim that every local species had been created where it is now found, and when Darwin sent Gray details of his theory, the latter seems to have absorbed it without too many qualms as ammunition for the fight. As an opponent of slavery, Gray also hated Agassiz's support for the anthropologists who declared the human races to be distinctly created species. It was partly because they shared these feelings that Darwin chose Gray as a confidant.[30]

In science, Gray emerges as a parallel to Hooker in Britain, yet on wider issues they were far apart. Gray was well primed to receive the theory of common descent and seems already to have been thinking along those lines when Darwin contacted him. Significantly, he was aware of Wallace's 1855 paper, referring to its law of species formation several times in his writings of the 1860s.[31] Common descent and dispersal were what mattered to him, and natural selection was only Darwin's supplement to those more important ideas. He struggled to reconcile the harshness and non-teleological aspects of natural selection with his religious beliefs. Arguing at first that any mechanism of adaptive species transformation was compatible with the belief that a wise and benevolent God governed nature, he was ultimately driven to accept that this argument did not work for natural selection. He suggested that God must have established the laws of variation to ensure that the production of new characteristics was purposeful rather than random, thereby erasing the need for the constant elimination of the unfit "scum of creation" implied by Darwin's theory.[32] Darwin protested that to suppose that variation itself was adaptive made selection superfluous, but in these passages we see how Gray, like Lyell and many others at the time, found it necessary to fudge the issue of how new characteristics were actually produced in order to retain a role for design by the Creator.

The evidence suggests that in a world without Darwin, Gray and Hooker would both have found their way to the theory of common descent in the 1860s. The trend of biogeographical discovery was pushing Lyell, Wallace, Hooker, and Gray toward a viewpoint whereby species were not fixed and endless modifications were produced as populations migrated around the globe. They were also aware that species and varieties faced the constant threat of extinction from the invasion of better-adapted rivals into their territory. They would thus have traced much of the path that Darwin himself followed in our own world twenty years earlier. Without Darwin they would have produced a theory of common descent, but there would have been much less focus on the actual process by which populations adapt and hence no theory of natural selection. For some, such as Hooker and possibly Wallace, the naturalistic methodology of Spencer's *Principles of Biology* would have encouraged the view that an organism could adjust to its environment via changed habits (for animals) or automatic physiological processes (in plants, but also in animals). This view would eventually congeal into a full-fledged naturalistic Lamarckism. But for more conservative thinkers such as Lyell and Gray, it was important to imagine that the Creator had programmed the laws of variation to achieve purposeful results.

Huxley and Owen

We can see similar moves toward evolutionism in the work of morphologists studying living forms through comparative anatomy, embryology, and fossils. This field was occupied by some of Darwin's leading supporters and opponents, most obviously T. H. Huxley and Richard Owen. Here the priority was to bring some sort of order to the bewildering array of structures found in living and extinct animals. The minute differences between species and varieties were of little interest, and while it was obvious that many of the superficial structures of species were adaptive, there was a widespread feeling that the laws governing the deep structural divisions within the animal kingdom represented something more fundamental than the sum total of endless local adaptations. The search for meaningful relationships among living forms had been going on from the start of the nineteenth century and had led some naturalists toward versions of evolutionism long before Darwin published. In the first half of the century those initiatives had often been dismissed as too speculative, but by 1859 the morphologists, like the biogeographers, were beginning to see a theory of common descent as the way forward. The morphologists had different priorities, however, and placed much less emphasis on topics like migration and local adaptation.

Morphologists often regarded groups of related species as having derived, in some sense, from a basic form or archetype that defined the essence of the group. Some linked the archetype to an idealist philosophy in which the basic pattern existed in the mind of the Creator, but recent historical studies have suggested that for others the archetype carried no such baggage—it was a practical way of trying to understand how the diversity of natural forms could be arranged into groups sharing common features. At this level, it would be easy to replace the archetype by a common ancestor from which the related species had diverged. The crucial question then became how that divergence takes place. In Darwin's theory, natural selection drives the adaptive trends generating divergence, but any process of modification could produce a similar effect, and it would not be necessary for all the divergent trends to be adaptive. Many morphologists were suspicious of the utilitarian view that all features of all species must have (or have had) an adaptive function, and so were willing to consider non-adaptive trends driven by internal biological forces. The problem with a theory based solely on local adaptation was the implication that evolution had no predictable trends, nothing that could be seen as a "law" of development. Natural selection merely reinforced this problem by using undirected ("random") individual variations as the raw material of adaptive change.

Even more divisive was the question of whether the processes driving modification through time were purely natural, or whether they might embody a teleological element, a divinely implanted tendency to change in a particular and ultimately meaningful direction. For naturalistic thinkers such as Huxley, the process had to be non-teleological, but it might still involve directing forces that were independent of the pressures of everyday life. For this reason Huxley doubted that natural selection could be the sole mechanism of evolution. But he was fiercely antagonistic toward Owen's more conservative position in which evolution, adaptive or not, could still be seen as part of a divinely preordained plan. Crucially, neither wanted to specify with any clarity the actual nature of the processes they envisioned. Owen was by no means the only naturalist who preferred to leave deliberately vague the whole issue of how "creation by law" actually worked. Trying to imagine how God's providence was built into the laws of nature was a challenge no one wanted to face. But even Huxley said very little about the source of the purely natural trends he seems to have envisioned in addition to the superficial activities of natural selection. It would be left for the biologists of the later nineteenth century to deal with these issues.

One consequence of these complexities is that the rival positions staked out in response to Darwin were by no means as clear-cut as popular histories imagine. Huxley supported Darwin but did not believe that adaptation, let alone natural selection, was the most crucial feature of evolution. He was enthusiastic about selection because it gave him a weapon in his struggle to establish a naturalistic worldview, not because he could use it in his science. Owen was not the archenemy of evolutionism depicted in popular accounts of the debate—on the contrary, his rejection of Darwin's theory rested on the hope of modernizing the concept of design by imagining preordained trends driving evolution in purposeful directions. These points are crucial for imagining how these figures would have behaved during the 1860s in a world without Darwin. If Huxley's enthusiasm for natural selection was rhetorical rather than scientific, would he have been so open to the suggestions flowing from Wallace, Spencer, and other advocates of evolution? Perhaps he would have retained his "a plague on both your houses" position much longer, only converting to evolutionism when Ernst Haeckel showed him how it could be used to throw light on the fossil record in the late 1860s. And if Huxley played a less active role in the emergence of evolutionism, would there be more room for Owen and the conservatives to promote the idea of a compromise in which evolution was the unfolding of a divine plan? In a world lacking the idea of natural selection, the divisions between the scientific naturalists and the conservatives would have been less clear-cut and the debates correspondingly less abrasive. The emergence of a scientific evolutionism would have been slower, but would have generated far less stress both among the scientists themselves and in the wider world.

Huxley's reputation as "Darwin's bulldog" has generated the popular assumption that he must have been an ardent enthusiast of natural selection. But since Michael Bartholomew's classic reassessment in 1975, historians have become increasingly willing to challenge this myth.[33] Huxley certainly defended Darwin against the critics who wanted to dismiss his theory out of hand, and he welcomed natural selection as evidence that new explanatory tools could be developed to bear on the question of the origin of species. Selection thus allowed him to abandon his agnostic position, but there is little evidence that he saw it as an adequate explanation of anything but the most superficial adaptive modifications. Selection *was* useful as a means of bating the defenders of natural theology, precisely because it explained adaptation without the remotest indication of divine planning. But in his early career especially, Huxley was not an "adaptationist" (to use a modern term)—he did not think that the complete internal structure of living things could be explained as a compilation of adaptive modifications. Like most morphologists, Huxley thought there were "laws of form" determining how structures develop in an organism, without reference to the demands of its environment. For this reason, Spencer's enthusiasm for Lamarckism would not have had much effect on his thinking, even in a world without Darwin. Spencer might eventually have been able to persuade Huxley that he should move toward a naturalistic program of evolution (he was doing this in the years before 1859, without much success). Wallace's work on biogeography did not impress Huxley at this stage, although he did take an interest in the topic later. Perhaps the debate with Owen over the ape-human relationship would eventually have forced him to think more carefully about whether "relationship" implied a common ancestor.

All of these factors built to a pressure that would force Huxley to face the fact, obvious to all of his contemporaries, that the naturalistic philosophy required an alternative to miraculous creation. If he wanted to show that the relationships between similar species had a natural explanation, the basic idea of descent from a common ancestor would have to replace the notion of the archetype, even if the latter was conceived in abstract rather

than idealistic terms. But how would the divergence of form be achieved—
what sort of natural processes would have appealed to Huxley's style of
biological thinking? There would certainly have been room for processes
that did not depend on the requirements of adaptation, except in the negative sense that new species with positively harmful characters would not be
able to perpetuate themselves. The sudden appearance of new characters
by saltation was certainly a possibility; in his review of the *Origin*, Huxley
chided Darwin for depending too much on the principle of continuity.[34]
More important was the suggestion that there might be laws controlling
variation, perhaps channeling it along predetermined lines. Huxley hinted
at this in his 1871 response to St. George Jackson Mivart's attack on Darwin and in another essay of 1878. There was no suggestion that such trends
had any teleological element, but if they were operative, they would produce a much more structured pattern of development than would natural
selection. These were ideas that Huxley might have expanded in a world
without Darwin, and they would have considerably reduced the gap between his thinking and that of his opponents.

The curious thing is that even with the example of Darwin before him,
Huxley complained that little was known about the laws of variation, but
he did little himself to investigate the topic. And he seems to have had no
desire to undertake experimental studies of how individual development
might be affected by forces generated within an organism. In a counterfactual world, he would reconstruct the laws governing the unfolding of living
forms solely from a study of the relationships between adult organisms.
This refusal to think about the actual causes of variation would also have
reduced the distance between his position and that of Owen, Lyell, and
the other thinkers who avoided the topic to leave room for the implication
that divine providence could somehow shape the direction of evolution.
At first he was not even prepared to use the ideas of common descent and
evolutionary trends to explore the fossil record. Huxley was opposed to
the widely accepted view that the history of life on earth was governed by
a progressive trend; one of the reasons he endorsed Darwin was because
natural selection did not imply inevitable progress. Even under the influence of Darwin, he made no move to use the general idea of evolution in
the palaeontological studies that were becoming an increasingly large part
of his technical scientific work. The reptile-bird intermediate *Archaeopteryx*

was described without reference to its evolutionary implications, and when he did move into this area, his focus was on birdlike dinosaurs.

The main influence that persuaded Huxley to begin the search for phylogenies (evolutionary lineages) was his reading of Ernst Haeckel's *Generelle Morphologie* of 1866. In a world without Darwin, it may only have been at this point that Huxley would begin to emerge as a spokesman for evolutionism.[35] His role in the early debates would have been much curtailed, and without the anti-teleological rhetoric generated by the selection theory, the link between evolutionism and aggressive philosophical naturalism would have been less obvious. Even so, it seems probable that by the 1870s Huxley would have become an evolutionist and would have used the theory both for scientific and rhetorical purposes. He would still have played a significant role in promoting the morphologists' project to reconstruct the history of life on earth from anatomical, embryological, and fossil evidence. This—not the investigation of variation and selection—was the main concern of evolutionists through the late nineteenth century, even in our own world. In a world without Darwin, it would have been the only game in town, and Huxley would have encouraged his disciples—comparative anatomists and zoologists such as W. K. Parker, George Rolleston, and E. Ray Lankester—to get involved. He would also have liaised with palaeontologists, including Othniel C. Marsh, who was unearthing evolutionary sequences from the fossil beds of the American West.[36]

In our world, Huxley used Darwinism as a weapon in his campaign to promote the interests of professional "men of science" (he still didn't like the word "scientist"). Highlighting the anti-teleological aspects of evolutionism served to make the remaining exponents of natural theology seem out of date, intellectually as well as professionally. But we tend to forget that there was a strong middle ground of scientists who retained some form of religious belief and were not so anxious to eliminate all traces of teleology. For them, the idea of "creation by law" or what became known as theistic evolutionism was a useful compromise. Several members of the Darwinian camp retained the hope that at least some aspects of the evolutionary process might only be explained by assuming some form of divinely imposed predisposition of life to vary in appropriate directions. We have already encountered this position in the thinking of Lyell, Gray, and even Wallace. Their position was an uncomfortable form of fence sitting, but there were others who wanted to forge the idea of designed evolution into a weapon against Darwinian naturalism. In a world without Darwin, and hence without natural selection to highlight the materialistic implications of evolutionism, these more conservative thinkers would be in a better position to offer themselves as leaders of the evolutionary movement.

The anatomist Richard Owen had been trying to create a space for this kind of compromise throughout the 1850s, against the opposition of even more conservative thinkers. Without the *Origin of Species* to stir up controversy, Owen would have been able to present himself as the leader of an evolutionary movement in the 1860s that could compete with Huxley's naturalistic philosophy on more equal terms. With Huxley playing a less active role, Owen could have realized his earlier ambition to create a scientifically viable yet still teleological form of evolutionism. It would seem more viable in part because the lack of focus on a materialistic mechanism in Huxley's camp would minimize the differences between them. Both schools of thought could engage in the effort to translate archetypes into common ancestors and to investigate evolutionary trends, agreeing to differ on the question of whether those trends might ultimately be explained in purely naturalistic terms. To some extent, this is what actually happened in our world, although the input from the conservative camp became lost to sight as our highly polarized image of the Darwinian revolution took shape. In a world without Darwinism, Owen and his supporters might have played a much more visible role.

Because he wrote a critical review of the *Origin*, Owen has frequently been depicted as antievolutionary. But his activities in the previous decade show that he himself was looking for non-miraculous explanations of the origin of species—the "question of questions," as he called in it his review.[37] Darwin even used Owen's descriptions of divergent adaptive trends seen in the fossil record as evidence. Previously, Owen had considered various non-evolutionary mechanisms, but like most other naturalists, he probably came to regard these as less plausible in the 1860s. Even without prompting by Darwin, he would have been anxious to take a lead in the less critical atmosphere that had emerged, and some form of evolutionism would be the best option for this. Admittedly, Owen was still uncomfortable with the idea of a close link between humans and apes, but in most other areas he was ready to explore the possibilities offered by the various concepts of evolution available. He may have been more willing than Huxley to acknowledge the role of adaptation, although he did not believe that all organic structure could be explained in utilitarian terms.

We know how Owen's thought would have developed, because he began to discuss the various possibilities in a technical paper written in the 1860s, and, given his previous activities, it cannot be argued that he was simply responding to Darwin. An 1868 paper on the dodo explicitly appealed to Lamarckism, but elsewhere Owen argued that many other species had characteristics that could not be explained in this way. Writing on the aye-aye in 1863, he dismissed a Lamarckian (or Darwinian) explanation of its tooth structure, but made the crucial move of arguing that all the lemurs could be derived from a common ancestral form. His alternative was what he called the "derivative" hypothesis, which he then developed in detail in the concluding chapter of his *Anatomy of the Vertebrates* in 1868. Here is Owen's definition of "derivation," which in our world contains an explicit contrast to Darwin's theory, indeed to any theory that explains evolution solely in response to environmental change: "Derivation holds that every species changes in time, by virtue of inherent tendencies thereto. 'Natural Selection' holds that no such change can take place without the influence of altered external circumstances. 'Derivation' sees among the effects of altered external circumstances, a manifestation of creative power in the variety and beauty of the results."[38] Substitute "Lamarckism" for "natural selection" and you have what Owen would have written in a world without Darwin. Using this approach, he went on to make important contributions to the understanding of major phases of evolution, including his recognition of the significance of the mammal-like reptiles discovered in the fossil beds of South Africa.

How was derivation actually supposed to work? Owen probably allowed for saltations, but the key factor was that the "inherent tendencies" to vary would push species consistently in a particular direction, whatever the environment. Sometimes the result might be adaptive anyway, but other trends would result in characteristics of no use to the species. This is what the next generation of evolutionists would call "orthogenesis." There might even develop characteristics useful to another species—Owen still accepted that old favorite of natural theology, the claim that the horse had been designed for humans to ride. Here was the crucial difference between his theory and anything acceptable to Huxley. Both believed that there might be built-in laws governing variation and forcing evolution into definite trends. But for Huxley, explanation by natural law had to exclude design, whereas for Owen the laws might have directions planned by their Creator. Huxley's trends had a purely natural origin in the internal constitution of the organisms; Owen's were at least in part supernaturally designed.

Apart from this (admittedly important) point, Huxley's and Owen's ideas on evolution were remarkably similar. Both argued for internally programmed trends, but they differed on how those trends were produced. This point resonates with the theme of a controversial article written by David Hull in 1975 in which he argued that the division between the Darwinian and anti-Darwinian camps was defined not by real differences of opinion but by personal and professional loyalties.[39] There were many naturalists who might easily have fallen into the opposite camp except for their professional contacts. This is less true of key figures such as Huxley and Owen, for whom the issue of teleology was really crucial. But many, including Lyell and the physiologist W. B. Carpenter, seem to have had sympathies with both sides. In a world without Darwin, such differences would have been even less visible, since Huxley would not have been pushing natural selection to attack the concept of design. Huxley and Owen would still have been bitter rivals, but it would be easier for others to float more loosely between the two camps, creating a more unified evolutionary position based on common descent, a Lamarckian explanation of local adaptation, and a substantial element of built-in developmental patterns.

The real debate would not have been between the supporters and opponents of design, but between the biogeographers, who believed that an open-ended evolution responded always to the local conditions, and those who preferred to think of evolution as having laws of its own, laws that predetermined its outcomes. When the astronomer Sir J. F. W. Herschel dismissed natural selection as "the law of higgledy-piggledy," he was responding to something more than the disquiet generated by Darwin's appeal to "random" variation as the raw material of evolution.[40] He was also voicing a widespread feeling that so important a process as the development of life on earth could not be explained as merely the sum total of an endless sequence of unpredictable migrations and trivial local adaptations. If even Huxley wanted laws directing variation along predetermined channels, we can appreciate just how important this feeling was. It was a conviction that might have united those who believed the predetermination was natural and those who still hoped to retain an element of design.

Haeckel and German Evolutionism

The expectation that there must be laws governing development was even more crucial in Germany. Here too there had been a concerted effort throughout the 1850s to throw light on the origin of new organic forms. Important advances had been made in recognizing that the fossil record showed a divergence of adaptive modifications in the course of geological time. H. G. Bronn, who would translate the *Origin* in our world, published a massive survey in 1858 that even included a diagram resembling a branching tree. But like Huxley, Bronn felt that the way forward was barred by the lack of empirical support for any natural explanation of how new forms developed. Also like Huxley, he saw Darwin's theory as an interesting hypothesis, but he remained unconvinced precisely because natural selection was not the kind of law-based explanation that would seem truly scientific by the standards of German philosophy. Here again the conviction that there must be predictable laws of development counted against the "higgledy-piggledy" of Darwinism or of any theory that reduced evolution to a chain of unrelated local adaptations. These reservations were plain in Bronn's translation of the *Origin* and tell us a great deal about how he and other German naturalists would have taken the debate forward in a world without Darwin.

Bronn himself would not have taken things much further, because he died in 1862. It was the much younger Ernst Haeckel who was inspired by Bronn's translation of Darwin to use evolutionism in his studies of the relationships between species and their development through time. Haeckel promoted what would become the iconic feature of late nineteenth-century evolutionism, the recapitulation theory in which the evolutionary history of a species could be traced in the sequence of forms, much like an individual develops as an embryo. He did this in a series of hugely popular writings (popular also in English translation), coupling the theory with his aggressive philosophy of "monism," which looked all too much like materialism to his opponents. Haeckel was certainly inspired by Darwin and, like Huxley, found the anti-teleological aspects of natural selection useful in his campaign against organized religion. But trying to imagine how his thought would have developed without input from the *Origin* leads us to a major disagreement among historians about just how deeply Darwin's selection theory became embedded in Haeckel's program. Was Haeckel a true follower of Darwin whose evolutionism reflected all those "higgledy-piggledy" aspects that the conservatives so despised? Or was he, like Huxley, someone who found natural selection useful as a rhetorical device in his public campaign, but made little use of it in his scientific work? In the latter case, perhaps Haeckel's emphasis on the progressive nature of evolution reflected a non-Darwinian way of thinking that, for all his protestations to the contrary, preserved some aspects of the old teleology. As the creator of the controversial term "pseudo-Darwinism" to describe Haeckel's way of thought, I stand very much with this latter interpretation, and this shapes my vision of how Haeckel's thinking would have developed in a world without Darwin.[41]

Bronn's translation reveals the reasons why German biologists were anxious to explore the idea of evolution and why they found some aspects of Darwin's theory hard to understand, let alone accept. Like many of his fellow naturalists, Bronn was interested in the pattern of development revealed by the fossil record, but also by the relationships between living structures. He knew that the record suggested branching trees of development and hence was primed to see the significance of a theory of common descent. The question was, did natural selection provide an adequate, law-like explanation of how the species diverged? Obviously the selection theory does depend on law-bound processes, but because those processes work only to bring about local adaptation, the theory ends up implying that the course of development is irregular and unpredictable, that is, higgledy-piggledy. Bronn's extensive comments on Darwin's text show that he found this aspect of the theory unacceptable.

Thanks to the influence of Alexander von Humboldt, German naturalists had deep interests in biogeography, but they focused on identifying clearly defined zoological and botanical regions, each with its own unique laws of development. They were not so interested in island biogeography—one of Darwin's chief lines of argument for the unpredictability of change. Bronn also found it hard to convey Darwin's discussions of animal breeding and artificial selection to his German readers, who had little experience of such things as pigeon and dog breeding. Here again, what was to Darwin a key line of evidence and a key explanatory tool did not easily transfer to a culture that did not share the interests of English gentlemen. Bronn's reaction to Darwin resembles Huxley's early agnosticism, and his desire to find laws of development parallels Huxley's hankering after laws governing variation that could predispose evolution along fixed channels. Unlike Huxley, though, he had no radical philosophical program that could benefit from the anti-teleological aspects of Darwin's theory, so he remained unwilling to offer real support.

It was precisely those anti-teleological implications that appealed to Ernst Haeckel as he looked for a theory that would underpin his own campaign against organized religion in Germany. Haeckel's reading of Darwin, in Bronn's translation, triggered an enthusiasm for evolutionism that meshed with his own efforts to understand the relationships between the Radiolarians, a group of invertebrates, in the early 1860s. Fritz Müller, another naturalist inspired by a reading of the *Origin*, would pioneer one of the chief tools that Haeckel would use in later reconstructions of evolutionary histories.[42] Müller was studying crustaceans and realized that the theory of common descent could explain how the various modern forms had diverged. He also showed that the early, larval stages of modern species threw light on what the common ancestor of the group had been like. This was the recapitulation theory, the idea that the development of a modern individual could be used as a model for understanding the evolutionary history of its group. Müller noted, though, that this would only be possible in cases where the new characteristics appearing in the course of evolution had been added to the existing pattern of embryological development. If variation was a distortion of, rather than an addition to, existing development, the older stages would be lost. There is no doubt that Müller saw all this as a vindication of Darwin's theory. He explicitly invoked natural selection to explain how different groups of crabs had adapted in different ways to living on dry land. But for most late nineteenth-century biologists, the recapitulation theory seemed to fit better with the Lamarckian mechanism of the inheritance of acquired characters, in which variation is not random because it is shaped by the adult's own efforts to adapt to its environment.

Müller obviously wouldn't have written his text in support of Darwin if the latter had not published the *Origin*, but it seems reasonable to suppose that he would have organized his ideas on how to reconstruct the history of the crustaceans in the course of the 1860s anyway, and in the absence of the selection theory, he might himself have opted for a Lamarckian explanation. The same pattern of events would equally have held true for Haeckel, given that he too was struggling to understand the relationships within the Radiolarians. These were naturalists inspired by the desire to see the

overall structure of natural relationships and keen to use embryology as a
clue to understand the much slower form of development that is evolution.
Where Bronn invoked the fossil record to show divergent patterns of adaptive modification, they could see immediately how to use this idea to throw
light on their own studies. In Haeckel's case, it was the variability of the
Radiolarians he studied, and the discovery of intermediate forms between
the species, that prepared him to appreciate the general theory of divergent
evolution expounded in the *Origin*.

There were other events in Haeckel's life that also shaped his response.
He came out in support of Darwin in 1863, but the death of his wife early in
the following year plunged him into despair and destroyed any remains of
belief in orthodox religion. He threw himself into his work and in 1866 produced his *Generelle Morphologie* in which he outlined his scheme for understanding the development of life on earth in evolutionary terms. Underlying
his scientific plans was a philosophy he called "monism," the idea that matter and mind are merely different aspects of the same universal substance.
Monism held that the visible universe was the only reality—there was no
wise and benevolent God behind the scenes shaping events. Like Huxley,
Haeckel welcomed the theory of natural selection because it helped him
to argue against the old teleology. He was happy to exploit the idea of the
struggle for existence, although he seems to have used it most often to explain the elimination of less successful varieties and species, rather than of
individuals in the same population. This has important implications when
we try to imagine how his ideas would have developed if there had been
no *Origin of Species* to trigger his conversion to evolutionism. Given the
nature of his scientific work and his commitment to monism, it seems likely
that he would have become an enthusiastic evolutionist in the course of the
1860s anyway. But how different would this evolutionism have been from
the Darwinism he promoted in our world?

Here we have to take sides in a debate raging among historians as to the
exact nature of Haeckel's Darwinism. Sander Gliboff and Robert J. Richards argue that Haeckel was a true Darwinian who fully appreciated the
basic message of Darwin's theory: evolution is open-ended and unpredictable, driven solely by the adaptation of species to changes in their local environment.[43] They concede that he believed it was inherently progressive
and that he invoked the Lamarckian theory of the inheritance of acquired
characteristics as well as natural selection—but Darwin too believed in

progress and also included a minor role for Lamarckism. Other historians, including Stephen Jay Gould, Michael Ruse, and myself, see major differences between Haeckel's project and Darwin's. We are struck by the extent of Haeckel's commitment to progress, which seems to go far beyond anything contemplated by Darwin. The latter accepted that most adaptations are not progressive in the long run—indeed, many are degenerative, as in the case of parasites. Only occasionally does evolution "invent" a new structure of general use in propelling life to a new level of development. For Haeckel, most variations are progressive—there was no sense that the vast majority of variants are useless or harmful and have to be eliminated. This is why Lamarckism loomed so much larger in his scheme (as it did, for similar reasons, in Herbert Spencer's philosophy of progress). Haeckel had only limited interest in topics such as island biogeography and animal breeding, cornerstones of the Darwinian vision of open-ended, undirected evolution.

For Haeckel, the main purpose of evolutionism was to reconstruct the history of life on earth, a project that Darwin refused to endorse except for specific case studies. Like Huxley and many other morphologists, Haeckel thought that this project ought to yield information on the laws governing animal forms. The "phylogenies"—a term that Haeckel coined—or lines of evolutionary descent filling the pages of his popular books show all lines advancing to higher levels of development. Haeckel did appreciate that there were many different ways in which the various branches of evolution could advance, but he was also happy to depict the whole evolution of life on earth as a progression toward the human form, all other types being mere side branches. This was a model that Darwin had deliberately avoided in the one diagram he included in the *Origin*, which has no "main line" of development.

Haeckel's use of the recapitulation theory, which he called the "biogenetic law," cemented the link between evolution and the purposeful development of the embryo toward maturity. Gliboff points out that Haeckel's use of this law did not imply the orthogenetic lines of development later postulated by other recapitulationists, most notably the American neo-Lamarckians (discussed in the following chapter). Like Darwin, he used embryos to throw light on ancient ancestral forms, not to imply that evolution was predetermined to advance in a particular direction. But the appeal to embryology as a key tool for the reconstruction of phylogenies helped to

PEDIGREE OF MAN.

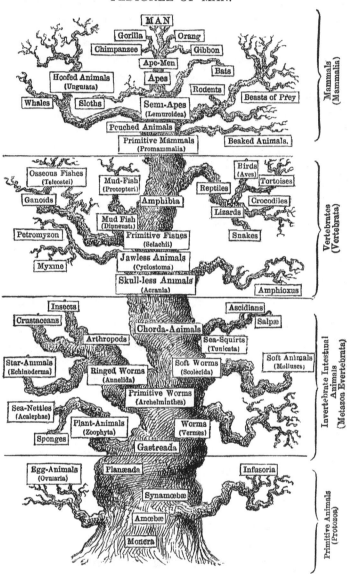

Figure 6 · Ernst Haeckel's tree of evolution, from his *History of Creation*. Haeckel produced many trees that did not have a main trunk leading upward toward humankind, but other evolutionists in the late nineteenth century endlessly copied this particular model.

create a very different impression of evolution, a "developmental" model in which life advances toward higher levels of organization almost automatically. This aspect of Haeckel's presentation of the theory gelled naturally with the model of linear development popular among cultural evolutionists. The white race became the pinnacle of biological evolution, just as European culture was the pinnacle of social progress. There was a built-in progressive trend, and although many side branches may have progressed too, it was only on the main line leading to civilized modern humanity that the full potential of evolutionary development had been realized.

All of this suggests that Haeckel's evolutionism was as much a philosophy of universal progress as a scientific evolutionism. It certainly wasn't a project to understand the everyday processes by which populations adapt to local conditions. Darwin spent much of his later life examining detailed cases of adaptation in the vegetable kingdom, Haeckel in promoting his broad evolutionary philosophy. For all Haeckel's distrust of religion, there remained an element of the old teleology in his assumption that the reaction of an organism and species to environmental challenge was normally positive and progressive. Other differences between Haeckel's thinking and Darwin's are also suggestive. Monism was often attacked as materialism, but in fact it held that mind and matter always went together so that there was a mental component in even the most basic material substances. In later years, this blurring of the distinction between mind and matter allowed monism to acquire some very odd bedfellows from within the ranks of esoteric and spiritualistic thinkers. Perhaps most striking of all, at least in the biological context, is the fact that Haeckel saw links between art and science. Where Darwin sought to explain away beauty as a product of sexual selection, Haeckel thought that life was inherently programmed to produce beautiful forms. One of Haeckel's last books was his *Kunstformen der Natur (Art-forms in nature)*, which depicted various species in forms inspired by the then fashionable art nouveau. Darwin's last book was on vegetable mold and the habits of earthworms. Could there be a clearer example of the differences between their approaches to the study of nature?

The point of this argument for the counterfactual history of evolutionism is that if Haeckel's thinking was not very Darwinian, it is easy to see how he would have moved toward a developmental model of evolutionism in the 1860s without being stimulated by the *Origin of Species*. This is not to deny that Darwin's book did kick-start his change of position, or that the idea of natural selection did not appeal to his emerging antipathy toward teleology. But the idea of natural selection acting on individual variants within a population was not crucial to Haeckel's thinking and could easily have been replaced more or less completely by Lamarckism. The application of the struggle for existence to explain the elimination of less successful species was something that was in the air anyway and was accepted even by anti-Darwinians such as Owen.

Without Darwin, Haeckel would have gone on to develop his progressionist evolutionism in the course of the 1860s, perhaps slightly more slowly, but in a form only slightly different from what we know. In the non-Darwinian world, Haeckel would still have produced his *History of Creation* and *Evolution of Man*. Those books would have stimulated scientists and the general public to appreciate the potential value of evolutionism for throwing light on the history of life on earth. But they would also have promoted a quasi-teleological form of progressionist thinking that paralleled the impact of Spencer in the English-speaking world. Both in German and in translation, Haeckel's works would still play a role in popularizing the idea that the main purpose of evolutionism was to reconstruct the history of life on earth with a view to seeing what lessons could be learned for human society and culture.

Haeckel's radicalism distressed many scientists who were willing to consider the general idea of evolution as the best way of understanding the development of life. Another radical naturalist, Karl Vogt, transferred his support from spontaneous development to evolution, but retained the formalists' vision of the history of life unfolding according to predetermined trends. Haeckel would not follow this lead, but his willingness to invoke environmental factors was constrained by enthusiasm for the idea of progress. He was convinced that the directing force of evolution must arise from a reaction of a living organism to changes in its environment. Any suggestion that forces from within prompted the change was dismissed as a concession to teleology. But as Huxley or Vogt might have pointed out, internally generated variation trends are only teleological if they are aimed in a purposeful direction; if they are merely a product of the organism's internal constitution, they have no such implication. Some of Haeckel's opponents, like Rudolf Virchow, paralleled Owen in retaining the idea that such internal forces were indeed purposeful; these scientists were the genuine targets of the radicals' ire. But if Haeckel had focused more on

Lamarckism, the contrast between his thinking and that of his opponents would have been less acute. In our world, Lamarckism became popular in the later nineteenth century precisely because it could be reconciled with teleology far more easily than the selection mechanism. Meanwhile, Vogt and other German naturalists were willing to parallel Huxley's preference for evolution driven by internal biological forces that did not necessarily imply design. These and other ideas would play an important role in shaping how evolutionary science would go forward in a world without Darwin.

5

A WORLD WITH A PURPOSE

Assuming that the scientific community of a world without Darwin still converted to evolutionism in the 1860s, how would evolutionary biology have developed over the next few decades? Which areas of investigation would have been most active? What resources would have been brought to bear on the project, and what ideas would have been used to explain how evolution worked? Most crucially, when would the theory of natural selection have been introduced?

It may seem difficult if not impossible to offer plausible answers to these questions. Projecting the course of counterfactual history becomes more difficult as we move decades beyond the point of divergence. But for the late nineteenth century, the complications are less serious than one might expect. We are not dealing with an alternate universe in which everything is different—this is not the equivalent of imagining an America in which the Confederacy won the Civil War, or a Britain conquered by Hitler. Taking Darwin's theory out of the picture has less effect because natural selection was not at first a very popular hypothesis. Most evolutionists in our own world did not believe that selection offered a satisfactory explanation, and many rejected it altogether or treated it as being of only marginal significance. It was a negative factor that weeded out the failures, not the source of genuinely new developments. Alternative hypotheses about how evolution might work were exploited so vigorously that by around 1900 many scientists believed that Darwin's theory was on its deathbed. The period was dubbed the "eclipse of Darwinism" by T. H. Huxley's grandson, Julian Huxley, in his mammoth 1942 survey of the theory that had been revived under the influence of genetics. If the selection theory ended up in such a parlous state in our own world, it should be possible to imagine how the alternatives would flourish in a world without Darwin.[1]

The alternatives were already in play in the 1860s. They include the Lamarckian theory of the inheritance of acquired characteristics, the idea that variation could be directed along fixed channels by internal biological forces (orthogenesis), and the theory of saltations or evolution by sudden jumps. More general resources included Herbert Spencer's increasingly popular evolutionary philosophy and the recapitulation theory promoted by Ernst Haeckel. Without Darwin's input, there might have been less attention focused on the actual mechanism of evolution. In our own world, many of the debates on that topic were prompted by a desire to discredit the selection theory. Without that motivation, scientists would have concentrated on trying to understand the results of evolution, not its causes. Histories of evolutionism focus on the debates over natural selection because hindsight tells us that Darwin's theory would eventually triumph. But that focus distorts our view of what mattered most to late nineteenth-century evolutionists. Far more effort was in fact devoted to reconstructing the history of life on earth from morphological, palaeontological, and biogeographical evidence. In a world without Darwin, that project would have been even more at the center of the stage.

The main tension would have been between a functionalist or adaptationist perspective that focused on environmental pressure as the main cause of change and a morphological approach seeking formalist laws governing the construction of living structures. Adaptationism was more Darwinian in the broader sense—it was committed to the theory of common descent and the utilitarian or functionalist assumption that all changes come about through species adapting to new environments. It would encourage the emergence of a theory of species selection in which species could be wiped out by more highly evolved forms invading their territory. But it was not tied to Darwin's theory of natural selection, and many anti-Darwinian biologists in the real world invoked Lamarckism to explain how adaptation occurred. The morphologists were less interested in superficial adaptations and more inclined to seek internal, biological forces that might generate trends in variation and hence in evolution. They were mostly associated with the rival formalist or structuralist position, and as they converted to evolutionism, their main concern was to understand the laws governing the development of new structures. There was also a formalist approach to biogeography that tried to identify regions governed by similar laws of development, but this was very different from the more dynamic and open-ended model used by the followers of Lyell and Wallace.

The formalists' preference for predetermined laws or trends created problems for the theory of common descent, because if two lineages were subject to the same trend, they might independently acquire the same characteristic—a similarity that those committed to the plasticity of variation would interpret as a relic of common ancestry. This is the concept of parallel evolution, and its significance is twofold. In the non-Darwinian world, this profoundly non-Darwinian concept of evolution might have had an even greater influence than it did on the evolutionism with which we are familiar. Breaking the hold of this model of predetermined development would have been even more difficult when the theory of natural selection did eventually emerge. Although biology in the world without Darwin might end up looking surprisingly similar to our own, there would be a significant difference of emphasis because the lack of the selection theory until the later nineteenth century would allow the idea of rigid evolutionary trends to gain an even greater hold.

In their different ways, both the adaptationist and formalist approaches encouraged the belief that evolution was an inherently progressive process and that by reconstructing the history of life on earth, a scientist could throw light on how that progress was achieved. Biogeography, allied with the Spencerian philosophy, presented progress as the end result of innumerable acts in which individuals or groups had solved the problem of how to conquer their environment. The formalist perspective encouraged a more structured progressionism that I call "developmentalism"—a tendency to see the development of the embryo as a model for evolutionary progression through stages of increasing maturity. In popular representations, this often meant a linear "chain of being" leading from the amoeba to humankind. Reconstructing the history of life on earth allowed biologists to define a hierarchy of development and gave them the ability to distinguish between the main line of advance and mere side branches that led to stagnation or degeneration. The adaptationist and formalist approaches both have implications for how biological evolutionism was used to endorse ideas of social evolution and the ideologies they represent. But with selection viewed as a mere culling of evolution's less successful efforts, there would not have been the acrimonious debates witnessed in our world as

Darwin's critics challenged his ideas about variation and heredity. This would be a world in which it was taken for granted that evolution was a purposeful activity aimed at pushing life upward along a morally significant scale of development.

In this scenario, evolutionism would flourish in various nonselectionist forms into the last decade of the century. There would be many similarities to the situation that played out in our own eclipse of Darwinism, although the developmental model would have enjoyed even more influence than it did, pushing the rival adaptationist perspective very much into second place. But when *would* the theory of natural selection finally emerge as a significant force? Given that natural selection and Lamarckism seem to be the only two conceivable mechanisms of adaptive evolution, it seems reasonable to suppose that the idea would have emerged once two conditions were satisfied. The first is a renewed focus on the assumption that evolution must have been shaped by environmental factors rather than by the purely biological trends favored by developmentalists. The second is a transition in thinking about heredity leading to a loss of confidence in the Lamarckian alternative. Neither of these conditions would be satisfied until the turn of the century.

THE NEW HISTORY OF EVOLUTIONISM

What would the history of evolutionism look like in the counterfactual non-Darwinian world? To imagine this, we can exploit new interpretations that have emerged among real-life historians of evolutionism in the last few decades. The field has been transformed since I became involved with it in the 1970s, and I have been privileged to play a role in that transformation. In the decades following the emergence of the modern synthetic theory of evolution—the neo-Darwinian synthesis of natural selection and genetics—the convention was to depict the history of the field as defined by a main line of development leading from Darwin to the synthesis. The focus was on the theory of natural selection, including the early debates about its plausibility and the way in which genetics solved the original problems. The title of the chapter on Gregor Mendel and the belated recognition of his experiments in Loren Eiseley's classic *Darwin's Century* of 1958 sums it up: "The Priest Who Held the Key to Evolution." The historical writings of Ernst Mayr, himself one of the founders of the modern synthetic theory, followed the same model. To the extent that historians acknowledged alternatives to the selection theory, they were dismissed as trivial blind alleys up which scientists were tempted, usually because they allowed their thinking to be influenced by their religious or moral beliefs.[2]

My own early work led me to realize the extent to which many palaeontologists of the post-Darwinian period had favored nonselectionist explanations of evolutionary trends, and I began the research that culminated in a 1983 book, *The Eclipse of Darwinism*. Even then, critics occasionally dismissed my work as a useful filling-in of a rather unimportant gap in the literature. But the more I worked on the non-Darwinian evolutionists, the more I became convinced that they were in the majority and represented the main thrust of nineteenth-century evolutionism. In my *Non-Darwinian Revolution* of 1987, I argued that it was Darwinism (as now defined) that was the real sideline until the early twentieth century; much of what passed for Darwinism in the previous decades owed little to what modern biologists see as Darwin's key insight. If we want to understand what really motivated the evolutionists of the earlier period, we need to look outside the boundaries defined by our modern concerns.

I then realized that to get fully under the skin of late nineteenth-century evolutionism, it was necessary to recognize that the debate over the mechanism of evolution was not the central issue as far as most biologists were concerned. The real focus of attention was the attempt to reconstruct the history of life on earth from morphological, fossil, and geographical evidence, a project that certainly had implications for ideas about the mechanism of evolution, but was not necessarily driven by that issue. The result was another book, *Life's Splendid Drama*, a title borrowed from the Canadian palaeontologist William Diller Matthew, who played a major role in several episodes of the reconstruction.

In recent years, this new interpretation of post-Darwinian evolutionism has been reinforced by other trends, most notably the recognition of the role played by the formalist alternative to creationism in German science. Nicolaas Rupke's account of the formalist tradition established by J. F. Blumenbach has revealed the origins of this approach to evolution. The belief that predictable, law-like processes governed the development of living structures provided the foundation for the developmental or non-Darwinian tradition. Historians of German science such as Rupke and

Wolf-Ernst Reif complain that this tradition has been largely written out of the history of evolutionism in the English-speaking world, thanks to the triumph of Darwinism in the mid-twentieth century (and the two world wars that were so damaging to German science and culture).[3] The new history of evolutionism rehabilitates this tradition and shows that it gained significant influence even in the English-speaking world in the late nineteenth century before being marginalized by the rise of neo-Darwinism.

This revisionist interpretation has become commonplace among historians of science and is slowly percolating through to the general community of scientists and historians. However, there is resistance. Michael Ruse, for instance, while admitting the role of non-Darwinian theories, still dismisses the project to reconstruct the course of life's history as second-rate science of no real importance to the overall emergence of modern Darwinism.[4] But for those whose main interest is to understand what motivated the first generation of evolutionists, this hindsight-driven assessment no longer seems appropriate. Fortunately, modern debates sparked by the emergence of evolutionary developmental biology have revived interest in the earlier developmentalism. Issues that fascinated the early evolutionists only to be dismissed by the neo-Darwinists might have some significance after all. If the attempt to understand the late nineteenth century on its own terms suggests that Darwin's theory should not occupy center stage (however important it might have been as an initial stimulus), then the alternatives that were called in to supplement or replace it provide a model for what we should expect to find in the non-Darwinian world.

Philosopher of science Ron Amundson is an enthusiast for evo-devo who argues that the traditional view of the history of evolutionism has tended to marginalize the contributions of topics and theories that, while losing popularity over the course of time, were once of major significance. His call echoes Rupke's appeal for us to rediscover the forgotten structuralist tradition in German biology. What I have called the developmental approach is characterized by these topics and theories, all reflecting the view that the role played by the development of the embryo must be acknowledged for a proper understanding of the evolution of living forms. Amundson tends to ignore some of the less successful aspects of the developmental tradition, for example, its tendency to promote extreme anti-adaptationist views such as the theories of orthogenesis and parallelism. Those exaggerated expressions of the analogy between individual development and the history of life on earth did have to be excluded for modern evolutionism to emerge. We should not allow the pendulum to swing too far; the earlier form of developmentalism did lead to some dead ends, and the need to bypass them should not in turn be written out of history. But Amundson is right to argue that neo-Darwinism threw the baby out with the bathwater.[5]

The exclusion of developmentalism led to the revival of Darwinism in the early twentieth century. In America especially, that process led to a divorce between genetics and embryology, promoting a version of neo-Darwinism that also paid little attention to how characteristics are actually formed by development. Evolution became just the shuffling of genes representing unit characteristics. Evo-devo emerged as evolutionists in the modern world recognized that this was an oversimplification. The pathways by which genes are translated into organisms are crucial to a true understanding of the forms that those organisms can express, but it took a good deal of controversy for this reemergence of development to become appreciated in modern biology. Amundson's point is that if we now recognize the importance of the embryo, we should not write histories of evolutionism that marginalize the extensive earlier interest in this relationship.

In a world where developmentalism became even more influential than it did in ours, the discovery of natural selection and its incorporation into evolutionary science might have been less disruptive. Instead of losing sight of developmentalism altogether, evolutionary biologists would have gradually abandoned its more extreme manifestations while at the same time incorporating the "new" theory of natural selection as a replacement for Lamarckism. We would end up with the same kind of rich synthesis of genetics, selectionism, and evo-devo that we have today, but the components would have been put together in a different sequence and the artificial oversimplification of Darwinism that characterized the mid-twentieth century in our world would have been avoided.

My counterfactual story of evolutionism in a world without Darwin suggests that Darwin's unique contribution gave support to the functionalist-adaptationist tradition, allowing it to steal a march on the originally much more influential developmentalism. It wasn't so much the early introduction of the idea of natural selection that was so crucial, because the theory was not successful at first. But Darwin's support for the biogeographical model of open-ended evolution allowed that model to gain a wider degree of influence than it might otherwise have had. It also established a framework within which the hereditarianism of early genetics was encouraged to abandon its saltationist roots. This created a foundation for the emergence of neo-Darwinism and the almost total eclipse of developmentalism in the mid-twentieth century. Without Darwin, the developmental approach would have retained much more influence, and we would never have viewed it as anything but a crucial component of the synthesis that gradually emerged. In the real world, developmentalism was marginalized (except in Germany), and we have had to recover some of its insights to create a more balanced view in which both the Darwinian and the formalist viewpoints have something to offer.

THE HISTORY OF LIFE

Evolutionists had three sources of information from which to reconstruct the history of life on earth: morphology, palaeontology, and biogeography. Each offered lines of evidence that could be used to throw light on how the first members of each new group had evolved from earlier ancestors, how the group had diverged into a complex of later, more specialized forms, and how the resulting species had migrated around the world to create the populations of the various regions. The two rival perspectives, functionalist and adaptationist, could be applied to each of these areas, although the nature of the evidence employed tended to bias scientists in one direction or the other. Morphologists tended to be functionalists, biogeographers favored the adaptationist approach, and palaeontology was a battleground that eventually tipped the balance toward an adaptationist perspective, permitting the emergence of a worldview resembling modern Darwinism.

Embryos and Ancestors

Morphology dominated the life sciences in the mid-nineteenth century, giving way only slowly to experimental studies of physiology and reproduction. Where field naturalists worked with live animals in their natural environment, morphologists studied internal structures through the dissection of preserved specimens in the laboratory. Increasingly the work also included detailed study of embryos, using ever more refined microscopical techniques. By looking for similarities between the internal structures of different species, the comparative anatomist and embryologist could hope to determine degrees of relationship, providing firm support for the groupings that field naturalists established on more superficial analysis. In some cases morphology overturned traditional views, as when the barnacles that Darwin studied turned out to be highly modified crustaceans rather than mollusks, as everyone had initially assumed. In the 1860s, morphologists began to realize that in determining such relationships, they were revealing not a collection of abstract "types" defining the groups, but evidence of divergent evolution from a common ancestor. As superficial modifications were added in the course of evolution, the more basic characteristics of a common ancestor would be preserved, allowing degrees of relationship to be established. A hypothesis could be proposed as to the nature of the common ancestor, while deeper relationships could be used to suggest the earlier form from which the founders of a new group had emerged.

We tend to assume that exploration of the fossil record was the primary source of information about the course of life's development on earth. But as Darwin realized, the record in the 1850s and '60s was still highly fragmentary and often suggested sudden changes rather than gradual evolution. The situation would change in later decades, at least for some groups and some transitions. But in the 1860s, the primary source of clues about ancestry still lay in the study of living things and their relationships. For the earliest stages in the history of life before there was any significant fossil record, morphology remained the only source of information. Ernst Haeckel proposed a hypothetical genealogy of the earliest animal forms, his Gastrea theory, which supposed that all animals had evolved from a simple spherical agglomeration of cells.

Here the recapitulation theory came in, because the best line of evidence for Haeckel's proposal was that all animals pass through such a stage in the earliest phase of their development from the fertilized ovum. But even for higher animals, as in the case of the barnacles, early embryonic stages frequently provided clues as to which main type a highly modified group had emerged from. The recapitulation theory suggested that the earliest stages—during which crustaceans look nothing like their adult stage—might actually correspond to the ancestral form of the group. The crustaceans provided the source of Fritz Müller's support for evolutionism. In Britain, Huxley's protégés E. Ray Lankester and Francis Balfour provided similar lines of evidence to establish the origins of the mollusks and the earliest fishes. Other morphologists gained crucial insights into later developments: Ernst Gaupp deduced the transformation of bones in the reptilian jaw into those of the mammalian ear even before the discovery of fossilized mammal-like reptiles confirmed the reality of the link.[6]

The process wasn't foolproof, however, and morphological studies alone could not resolve some important problems. The classic example was the ancestry of the vertebrates (chordates). Haeckel and his followers noted a resemblance between a primitive chordate, the lancelet, and the tadpole-like larvae of sea squirts, or ascidians. They suggested that this group was the ancestor of the vertebrates, with modern sea squirts (which are sessile creatures that cling to rocks) being degenerate offshoots of the vertebrate stock. But Haeckel's rival Anton Dohrn argued that this link was implausible because it ignored the fact that the vertebrates, unlike the tunicates, are segmented animals. He proposed that vertebrates were in fact arthropods that had somehow been turned on their backs to reverse the position of some key structures. This made segmentation the more primitive characteristic, linking vertebrates with groups such as insects.

The problem was that there was no criterion by which one could unambiguously determine which characteristics were truly primitive (and hence ancestral) and which were later-derived modifications. Evolutionists accepted that in some cases it was possible for two groups independently to evolve very similar characters. This could occur either by convergent evolution, when two lines adapted to very similar lifestyles, or by parallelism, whereby the same built-in variation trend affected the two types. In disputed cases, one side's primitive character was the other's parallelism and vice versa, and in the absence of fossil evidence, there was no way of telling who was right. The controversy became so scandalous by the end of the century that evolutionary morphology was discredited, encouraging the transition to experimental work. William Bateson, for instance, gave up his study of the ancestry of the vertebrates to help found the science of genetics.

This situation is the source of Ruse's claim that the work of this generation of evolutionists was second-rate science, although they could hardly have been expected to realize the limitations of their techniques when the project began (some of the problems have recently been resolved using DNA evidence). It is possible to take a more sympathetic view of their efforts. Partial resolution of the vertebrate scandal was in fact provided around 1900 by the embryologist E. W. MacBride, who recognized the underlying similarities between vertebrates and echinoderms (sea urchins and starfish). MacBride was one of the last great exponents of the recapitulation theory and a convinced Lamarckian, proving that Darwinism was not essential for serious work in this area. None of the morphological techniques depended on a detailed understanding of the theory of natural selection, although the question of the extent to which evolution was driven by adaptation or predetermined trends was certainly important. Those who supported Lamarckism and orthogenesis tended to believe that there would be more cases of parallelism than the Darwinians would allow. Some Lamarckians abandoned the adaptationist position to argue that evolution was driven by predetermined trends. One difference between the non-Darwinian world and our own would almost certainly be that without the influence of Darwinism, there would have been even more enthusiasm for the idea of predetermined evolution.

The Fossil Record

Darwin was pessimistic about using the fossil record to trace the course of evolution, but new discoveries were being made all the time, and the situation soon changed dramatically. An evolutionary movement would almost certainly have started by the 1870s even without the impact of the *Origin of Species* because the flood of new fossils began to make the theory of gradual development far more plausible. Trends in the history of certain groups were already apparent by 1860, as Owen had shown. As more fossils were discovered, the gaps in some of the sequences were gradually reduced to the point where it seemed easier to imagine evolutionary links than stick with the old idea of a related sequence of miracles. Equally importantly, crucial fossils showing links between the major vertebrate groups were found, undermining the view that each class had a totally separate origin. Even today there are only a few of these transitions that can be charted all the way through with hard evidence. But as each intermediate is found, the creationists' insistence that all major groups have a distinct origin has to be further qualified. Quite early on, their position became open to the claim that if the Creator wanted to keep the groups totally isolated, He was remarkably careless about what He was doing around the edges. In at least one case, the transition from reptiles to mammals, a continuous sequence of fossils showing the intermediate stages was available by the 1890s.[7]

By the end of the century, palaeontology was beginning to take over from morphology as the chief focus of attention for evolutionists seeking to understand the overall history of life on earth. There were also important developments linking the appearance of new types to events in the earth's physical history. As a result of all this activity, early twentieth-century surveys of the process have a much more "modern" feel to them than the rather abstract accounts of the relationships between types offered in Haeckel's popularizations from the 1870s. None of this was a direct consequence of Darwin's particular theory of evolution, and it is plausible to imagine that it would all have taken place in a world where he had not been a participant. There were, however, differences of emphasis between Darwinian and non-Darwinian palaeontologists in our world, so as with morphology, it is possible that formalist interpretations of how evolution worked would have had a freer rein in the counterfactual world.

The most crucial discoveries were those plugging major gaps in the fossil record, especially the origin of new vertebrate classes. Most attention has focused on the transition from reptiles to birds, a study in which the discovery of the first specimen of *Archaeopteryx* in 1861 provided a fossil that could easily be seen as transitional. Curiously, neither Huxley nor Owen described it in this context at first, and even when more specimens were found, *Archaeopteryx* remained something of an anomaly. In 1868, Huxley, now a convert to using the fossil record for evolutionary purposes thanks to Haeckel, wrote a paper on the animals intermediate between birds and reptiles.[8] Here he referred to *Archaeopteryx* but then continued on to discuss the birdlike feet and legs of some dinosaurs. He used the term "missing link" to describe these fossils, although what was really needed was a series of links demonstrating the whole sequence of development. Huxley theorized that birds had evolved via an ostrich-like stage, with flight coming later, an idea soon challenged by an opposing theory that supposed that flight came first through forms resembling *Archaeopteryx* (although the latter was too late in the record to be the actual ancestor). The story of bird origins, like that of other classes, turned out to be far more complex than the early evolutionists assumed, with the available fossils being too scattered to allow a decision to be made between opposing theories. Even so, the fossils were enough to break down the creationists' claim that birds appeared without any sign of transition from a previous class.

This story was repeated in the case of the transition from fish to amphibians, where some important fossils showed that fish were acquiring limb-like fins before the move onto land began. But debates continued as to whether the modern lungfish were remnants of a transitional form, or whether the lobe-finned Crossoptygian fish were more important. There was even a suggestion that modern frogs and newts had two independent origins in different ancestral classes of fishes. Here we see the power of the concept of parallel evolution to undermine the credibility of the theory of common descent by suggesting that similar evolutionary trends might produce the same result in two independent lineages. Exactly this point was made in the case of the mammals by the anti-Darwinian morphologist St. George Jackson Mivart. He argued that the monotremes of modern Australia (such as the egg-laying duckbilled platypus) had evolved some mammalian characters independently of the line leading toward the now dominant placentals. The conviction that there were laws of development pushing evolution in predetermined directions thus functioned very actively even in our own world. In a world without Darwin, it would have been possible for such ideas to have flourished with even less opposition.

In the case of the mammals, the discovery of the mammal-like reptiles of the Permian rocks of South Africa eventually clarified the process by which the complex jaw structure of the reptiles was transformed into the simple jaw and the minute ear bones of mammals. Significantly, Richard Owen first noticed the potential importance of these fossils, and Huxley pointedly ignored them. It was a disciple of Owen, Harry Govier Seeley, who made some of the first detailed studies beginning in the late 1880s. In 1932, Robert Broom, now better known for his work on hominid fossils, wrote the most comprehensive account of the transition from reptile to mammals. Broom was a profoundly non-Darwinian evolutionist who remained convinced that the progress of life on earth was the unfolding of a divine plan aimed at the production of humankind.[9] Here we see a case of major innovations being made by naturalists openly hostile not just to natural selection but to the whole materialist worldview promoted by Huxley and his allies. It cannot be argued that Darwinism was the only theoretical perspective from which serious studies of evolution can be made, and in a non-Darwinian world, the teleological, developmental view would have been even more active in shaping efforts to understand the history of life on earth.

The same point can be made for another area when fossil discoveries aided the evolutionists' cause. By the 1870s, discoveries from various parts of the world were beginning to fill in the details of how many modern (and some extinct) groups of animals had evolved. The classic example was that of the horse, for which a whole series of fossil antecedents was discovered in the rocks of North America in the 1870s. By 1877, Huxley was able to proclaim the series of horse fossils described by Othniel Charles Marsh as "demonstrative evidence of evolution"—and this was before the discovery of the most primitive form, the little *Eohippus* that showed how the horse family had evolved from small woodland grazers.[10] Evolutionists could also trace the rise and fall of numerous now extinct families, including several giant mammalian types, for example, Marsh's Dinocerata and the Titanotheres found in the rocks of several Western states and later described by Henry Fairfield Osborn. Critics of evolution complained then as now that the sequences were not completely continuous, but they were enough to make it plain that if the Creator was starting each species afresh, He often worked to a remarkably consistent pattern.

Marsh and Huxley shared a naturalistic view of evolution, but the trends revealed in the evolution of many groups were widely interpreted as evidence in favor of developmental constraints that pushed the species inexorably toward a preordained goal. In Germany, Albert von Kölliker argued that if developmental laws controlled variation, there was no reason why the same species could not be developed independently in lineages sharing the same trend. This would undermine the theory of common descent, in which the underlying similarities on which relationships are based indicate structures retained from a common ancestor. For many late nineteenth-century palaeontologists, the trends they saw in the fossil record were just too regular to be the product of a haphazard process at the mercy of every environmental fluctuation. Members of the American school of neo-Lamarckism were leading champions of this idea, and in our world they became outright opponents of the Darwinian selection theory. Edward Drinker Cope, working with vertebrate fossils, and Alpheus Hyatt, working with invertebrates, both insisted that they saw rigidly parallel lines of development within the groups they studied. Some mysterious force was

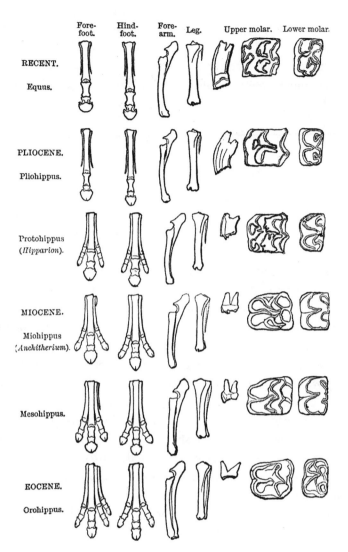

	Fore-foot.	Hind-foot.	Fore-arm.	Leg.	Upper molar.	Lower molar.

RECENT.

Equus.

PLIOCENE.

Pliohippus.

Protohippus
(*Hipparion*).

MIOCENE.

Miohippus
(*Anchitherium*).

Mesohippus.

EOCENE.

Orohippus.

Figure 7 · Fossils that T. H. Huxley used to illustrate the evolution of the horse and explain the new discoveries made by O. C. Marsh, from "Lectures on Evolution," in Huxley, *American Addresses*. Note how easily the simple arrangement of a few fossils gives the impression of a straight line of development toward increasing specialization.

directing variation along predetermined lines, and this was strong enough to produce more or less identical species in separate lines of evolution. Cope's disciple Henry Fairfield Osborn continued to develop this idea well into the twentieth century, using evidence from the extinct mammalian families he studied.

Biogeography

The most obvious counterweight to the developmental view of evolution was the adaptationist model that—even without Darwin's influence—would have been explored through the later nineteenth century. The theory of natural selection was not the only vehicle that could promote the idea that evolution was driven mainly by unpredictable environmental stresses. Darwin used biogeographical evidence, including that from the Galapagos Islands, to illustrate the divergence of a group of related species from a common ancestor when an original population is broken up by geographical barriers and exposed to different conditions. Wallace independently hit on the same clue even before he began to think in terms of one variety actually replacing another because it was better adapted. In a world without Darwin, Wallace, Hooker, and Gray would have moved toward a theory of divergent evolution driven by migration, local adaptation, and extinction. All that they needed was the basic idea that local adaptation shapes the changes within populations; the sheer unpredictability of migration and environmental change then generates a model of haphazardly branching evolution. Without Darwin, this move would have taken longer and would not have involved the theory of natural selection acting on individual variants. But it would certainly have emerged as an alternative to the developmental vision of evolution predetermined by law.

Darwin used biogeographical examples as case studies to support the plausibility of divergent evolution but made no effort to reconstruct the global pattern of life's history. As more evidence accumulated from around the world, including from the fossil record, it became feasible to attempt such a reconstruction. This would provide a comprehensive alternative to the developmental view of the history of life on earth. It was Wallace, not Darwin, who eventually launched this project. Although initially doubtful about the availability of information from around the world, in the 1870s he began to pull all the evidence together and in 1876 published his two-volume *Geographical Distribution of Animals*. By linking modern distributions with fossil evidence, he was able to work out the likely home territory where each major group had first appeared and then trace its subsequent migrations. His book triggered an explosion of research in the field through the later decades of the century, and there is no reason to suppose that this initiative would not have been taken even in a world without Darwin.[11]

Wallace showed how the modern distribution of animal groups could be understood by reconstructing the history of their migrations. The tapirs, for instance, now live as two isolated populations in South America and the East Indies. By appealing to fossil evidence, Wallace argued that the group had probably appeared in northern Eurasia, from where it had migrated southward to the East Indies and across the Bering Strait to North and eventually South America. Over most of that range it has subsequently been driven to extinction by the invasion of better-adapted types, leaving the two relic populations in the modern world, protected by their isolation. The migration of species to new territories was sometimes made possible by changing conditions, especially the lowering of the sea level during the ice age. At such times animals could cross between Eurasia and North America, and could spread from continents to occupy what have now become offshore islands. The division between the Eurasian and Australian faunas in the islands of the Malay Archipelago, which became known as Wallace's line, was marked by a deep channel between Bali and Lombok that, even when the sea level dropped, had never been bridged.

Wallace believed that the continents had occupied their current positions at least through the Tertiary period (the Cainozoic). In the absence of a theory of continental drift, he had to postulate accidental transportation by storms and vegetation rafts to explain how some species had migrated across the open ocean. Some biogeographers postulated the temporary emergence of "land bridges" across the deep oceans at some points in the past, although the geological evidence increasingly began to make earth movements on this scale seem implausible. The one thing that most naturalists agreed on was that the main center of progressive evolution was the northern regions of the earth, especially northern Eurasia. Here was the largest and most varied territory, with the most challenging environmental conditions. These pressures had encouraged the appearance of successively more advanced animal types, each of which subsequently migrated southward, often exterminating the previous occupants in the process.

This general principle was used to explain the distribution of the tapirs and many other groups. It was a model of evolution that already began to stress the episodic rather than gradual nature of the process—periodic waves of migration followed by extinctions. The scope of biogeography was also expanding thanks to knowledge of the distribution of species in earlier geological epochs. Eventually this knowledge would provide a real threat to the rival model of evolution based on parallelism and internally driven trends.

The biogeographical model of evolution cried out to be described in language paralleling that of Western imperialism.[12] Even Wallace, who was a socialist and no enthusiast for European expansionism, occasionally used terms such as "invasion" and "colonization" to describe the process by which more highly evolved types migrated into new territories and caused the marginalization or even extinction of the previous occupants. Most of the naturalists who subsequently worked in the field used such language quite unconsciously, and not surprisingly this aspect of evolutionary science was then used to justify the Europeans' occupation of territories around the globe as a natural extension of the evolutionary process. This is a relationship often described as a key feature of social Darwinism—yet it does not depend on the theory of natural selection acting within populations. An ideology of struggle and conquest could draw inspiration from the biogeographical evidence for migration and extinction, with the Malthusian image of a struggle for existence being applied at the level of species and territories rather than individuals.

THEORIES AND THEIR IMPLICATIONS

In the absence of Darwinism, several alternative theories would be available to explain how evolution worked. Each had particular implications for naturalists seeking to understand the history of life on earth. The differences between them were crucial for shaping rival evolutionary worldviews, but there were also underlying assumptions accepted by almost everyone in the field. In our world, Darwinism tended to get absorbed into evolutionary cosmologies with implications that Darwin himself might not have endorsed. In particular, most late nineteenth-century evolutionists took it for granted that the history of life was essentially progressive. Darwin himself accepted the idea of progress, but in a highly qualified manner. For him the upward trend was haphazard and irregular; most branches of the tree of life were not progressive in any long-term sense, leading only to specialization for a particular way of life. Breakthroughs to new levels of organization occurred only sporadically. Many of his contemporaries preferred to believe that progress was inevitable, something built into the basic laws driving the evolutionary process. Even when they looked at trends toward specialization, they saw them as something inherent, not just as byproducts of the day-to-day pressures of local adaptation.

The two major sources of this developmental model of evolution were the progressionist cosmology of Herbert Spencer and the analogy between embryology and evolution typified by the recapitulation theory of Ernst Haeckel. It would be hard to overestimate the influence of Spencer on the evolutionary movement in Britain and especially in America (nor was he completely ignored in Continental Europe). For those more interested in social evolution than in natural history, it was Spencer rather than Darwin who served as the figurehead for the movement. Many scientists too were impressed by the overarching evolutionary philosophy that Spencer proposed as a framework for their work.[13] Several of the figures we associate with Darwinism and its revival fall into this category, including the American biologist Sewall Wright. Spencer's was a naturalistic philosophy that denied that any indication of supernatural design could be seen in nature, although Spencer postulated an "Unknowable" lying beyond the material universe. For him, the laws governing the material world ensured a steady trend from homogeneity to heterogeneity, that is, toward increasingly more complex structures. He saw complex living bodies as analogous to a specialized modern society, with each organ doing its own job for the good of the whole.

Spencer didn't stress the analogy with embryology, so his philosophy was somewhat less developmental than Haeckel's. For him, evolution was the summary of innumerable interactions between individuals and their environment, and to this extent his perspective was closer to the adaptationist viewpoint we associate with Darwin. Degeneration was possible if a group of organisms began adapting to a less active way of life. But the overall trend of most lines of evolution was upward, even if their paths were somewhat ragged. In principle, this was a vision of branching evolution with no single goal, since there were many different ways of becoming more complex. But Spencer's followers all knew that one line in particular led to the most important breakthrough of all: the human mind and a new phase of social and cultural development. Most cultural evolutionists assumed that there was a main line of development leading from primitive savages through to modern industrial civilization.

Although deriving from Germanic rather than Anglophone sources, the monistic philosophy that Haeckel associated with his evolutionism paralleled the Spencerian cosmology. Monism held that mind and matter were opposite manifestations of a single underlying substance, thus undermining the charge that it was materialistic. But the notion that even material atoms had a rudimentary consciousness left the movement open to links with esoteric philosophies and certainly encouraged the view that purposefulness was somehow built into nature.[14] Like Spencer, Haeckel stressed the progressive nature of evolution, with the assumption that embryological development served as a record of a species' history helping to give the impression that both were somehow directed toward a goal. In principle, Haeckel's progressionism was also divergent, each branch of the tree of life progressing in its own way. The reaction between an organism and its environment was crucial to eliciting progressive variation. But popular accounts tended to treat the line leading toward the human species as more significant, all the other animal groups being mere side branches. To those who saw embryology as the key to evolution, even the side branches had their own built-in trends. Whatever Haeckel's own insistence that internal, purely biological forces could not lead to progress, many supporters of the recapitulation theory held that such internal factors were powerful enough to drive variation in a single predetermined direction.

Lamarckism

Another similarity between Spencer and Haeckel's evolutionism is that both were strongly wedded to the Lamarckian mechanism of the inheritance of acquired characteristics. Even in our world, both thought that Lamarckism was more important than natural selection operating on random variation—indeed, Spencer later emerged as a leading critic of neo-Darwinism. In a world without Darwin, both men would have promoted an essentially Lamarckian evolutionism and would have encouraged others to follow their lead. The Lamarckian theory went far beyond the idea of "use-inheritance," often caricatured in the "just-so" story of how the giraffe got its long neck. The reference to Rudyard Kipling's famous stories highlights the extent to which this image has influenced our culture, often without anyone realizing that it reflects the Lamarckian version of evolution. In Kipling's stories, and in the popular example of the giraffe, a species changes because individual animals change their behavior, exercise their bodies in different ways, and thus develop the affected parts.[15] The ancestral giraffes adopted the habit of feeding from trees, perhaps because of a change in their environment, and so stretched their necks upward. Their necks became slightly longer as a result of the exercise, and—this is the crucial assumption—their offspring inherited the extension and continued the process because they too stretched their necks. In this concept of use-inheritance, the animals' deliberate choice of a new habit shapes their bodies, and the results accumulate to generate an evolutionary modification of the species.

This is the best-known version of Lamarckism, although the giraffe example isn't the best illustration of how exercise might change animals' bodies. Think instead of the weightlifter's enlarged muscles and imagine what might happen over many generations if such changes really were inherited. But the theory is actually more sophisticated than this. Lamarck himself recognized that use-inheritance cannot work for plants (because they can't change their habits) and proposed instead that their internal structure could spontaneously respond in a purposeful way to stresses imposed by the environment. If a plant grows in dryer conditions, it develops a thicker external covering to retain water. The Lamarckians of the later nineteenth century realized that this process could work with animals too: an animal raised in a colder environment might grow thicker fur, even though this is something over which it has no conscious control. Both Spencer and Haeckel envisioned this kind of spontaneously purposeful variation occurring within living bodies, and both took it for granted that it would be inherited, along with the effects of deliberate exercise. Until the emergence of the modern conception of heredity in the last decades of the nineteenth century, almost everyone (including Darwin) thought there would be interactions between the body and its reproductive system allowing any useful modifications to be passed on to the next generation.[16]

Acceptance of the inheritance of characteristics acquired by an individual organism was so prevalent that few thought it necessary to test the effect experimentally. When rigorous tests were applied later in the century, it proved difficult to find confirmation of the effect, and opinion began to shift toward the modern view that there is no process by which changes in a parent's body can be transmitted to its offspring. But in the 1860s and '70s—even in our own world—few Lamarckians thought it necessary to defend their theory with experimental proof. Indirect arguments supposedly confirming the effect were sufficient. In some cases, simply demonstrating that parent and offspring both acquired the same characteristic was deemed to be sufficient proof, even when the offspring was exposed to the same conditions and could thus have acquired the new characteristic independently. In a world without Darwin, where there would have been no controversy over the selection of random variations, there would have been even less need to focus attention on the experimental proof of Lamarckism or the details of the mechanism of transmission.

In its broadest form, Lamarckism was a flexible theory that could be used in several ways by scientists seeking to understand how life evolved. In its most basic form, it could be conceived as a simple process of adaptation, fulfilling much the same function as natural selection did in Darwin's thinking. Both Spencer and Haeckel presumed that modifications occurred in response to environmental change and that the results were adaptive. Those naturalists and biogeographers who focused on migration and local adaptation as the driving force of evolution could thus exploit Lamarckism. The theory remained popular in some fields well into the twentieth century. Ernst Mayr, one of the founders of modern neo-Darwinism, recalled that when he began his career as a field naturalist studying the birds of the Indonesian islands, his training in Germany led him to take it for granted that local adaptation was the result of Lamarckism. It was only later that he began to appreciate that developments in genetics were undermining the plausibility of the theory he had been taught.[17] Recent research shows that some British field naturalists were sympathetic to Paul Kammerer's Lamarckism when he visited the country in 1923.[18] In the world without Darwin, the plausibility of Lamarckism would have seemed even more unassailable, at least until other factors began to focus attention on the process of inheritance.

The presumption was that most changes were a positive response to the environment so that evolution was normally progressive. Most cases of adaptation entailed increased specialization for a particular way of life, as with the giraffe's long neck, so in its simplest form, Lamarckism could support the theory of divergent evolution and the principle of common descent. Scientists also recognized that at least some adaptations were degenerative rather than progressive. An example widely cited as indirect evidence in favor of use-inheritance was the blind animals found in caves. Naturalists such as Alpheus Packard in America studied these creatures and were impressed by the fact that these now useless structures had degenerated to the extent that they had disappeared altogether. Some Lamarckians argued that Darwinism could not explain this kind of adaptation—why should natural selection favor the complete elimination of a structure that had been essential in the recent past? In the non-Darwinian world, it would be taken completely for granted that Lamarckism was the only reasonable explanation.

This example points us to one of the resources used by writers who applied the evolutionary perspective to social issues. For Spencer and his followers, a key aspect of the theory's implications for human behavior was that in a competitive environment, the system rewarded those individuals who displayed effort and initiative, but it penalized laziness and stupidity. In his popular book *The Water Babies*, clergyman Charles Kingsley stressed how those who chose to be active and resourceful were rewarded with progress to a higher state, while those who were inactive would degenerate. At the level of species, this model could be adopted by Darwinians. Huxley's protégé, the zoologist E. Ray Lankester, wrote a book entitled *Degeneration: A Chapter in Darwinism* in which he noted that species that adapted to a sessile lifestyle invariably degenerated. He used the example of the tunicates or sea squirts from which the vertebrates were thought to have descended. Their tadpoles were still active creatures, but the adults had degenerated into fixed organisms attached to the seabed. In our world, Lankester offered this as an illustration of Darwinism: species adapting to a sessile lifestyle would experience natural selection adapting them to a life that did not require the organs of locomotion. But Lankester himself was at first tempted by the Lamarckian element in Haeckel's evolutionism—he had supervised the English translation of *The History of Creation*. For most of his contemporaries, the degenerative modifications were presumed to begin at the individual level, and it was the consequences of the organisms' own efforts (or lack thereof) that shaped the future of the species via the Lamarckian effect.

Lamarckism also turned out to be flexible in terms of its moral and religious implications. As part of Spencer's philosophy, it was associated with scientific naturalism and materialism. It was also a key aspect of Spencer's liberal and individualist social philosophy—the struggle for existence within a population of competing individuals was the best stimulus to encourage effort and initiative in order to avoid the penalties of failure. The resulting effects of self-improvement would accumulate through inheritance to improve the quality of the whole race. Many liberal religious thinkers welcomed Spencer's focus on thrift and industry, although they placed less emphasis on the idea of competition.

Later in the century, Lamarckism was transformed into a very different worldview actively opposed to the materialism associated with Spencer's cosmic progressionism. If we think of the process by which the giraffe got its long neck, the ancestors who began the trend had to adopt a new lifestyle based on feeding from trees. This could be presented as an active choice, a sign of initiative and creativity that would determine the whole future of the species. By focusing on this element of choice, Lamarckism could be linked to a philosophy highlighting the ability of living things to make decisions and take control of their future. Instead of a mechanistic world where things happen in accordance with rigid laws, we have a vitalist philosophy in which life becomes an active purposeful force, perhaps even an expression of the Creator's powers delegated to living things.

We associate this vitalist Lamarckism with the novelist Samuel Butler's attack on Darwinism in his *Evolution Old and New* of 1872.[19] Originally a supporter of Darwin, Butler was inspired by Mivart's assault on the selection theory and began to see Lamarckism as a means of attacking what he perceived as a soulless materialism based on chance and death. In Darwin's theory, the organism was powerless—it lived or died according to the success of the random variations it had inherited. Lamarckism empowered the organism and raised life above the material world. The Darwinians ostracized Butler, but his views later became influential even among scientists—including Darwin's son, Francis. In America, the Quaker palaeontologist Edward Drinker Cope, a founder of the neo-Lamarckian school, wrote a *Theology of Evolution* in 1887 that stressed the compatibility of the theory with a religion in which God's power was active in nature. He postulated a non-materialistic growth force called "bathmism" as the source of an organism's vital powers. A similar attack on mechanism associated with Henri Bergson's philosophy of creative evolution became a powerful influence on many early twentieth-century biologists including Julian Huxley.

Butler's target was the theory of natural selection, which seemed to encapsulate all the worst aspects of the mechanistic worldview he hated—it was a "nightmare of waste and death."[20] In a world without Darwin, there would have been no selection theory for him to attack, but the mechanistic worldview would still have been present through the scientific naturalism of Spencer and Huxley. The transition to a vitalist form of Lamarckism would almost certainly have taken place, with Spencer being the principal villain suffering Butler's invective. Perhaps the language might not have been quite so vitriolic, since this would be a transformation within the Lamarckian camp rather than an attack on an even more visible manifestation of materialism. But the rise of vitalism, with its associated religious implications, seems to represent a major shift in late nineteenth-century thought that would have occurred whether or not Darwinism was present to symbolize the materialism of the 1860s and '70s. One consequence of the transition for our own perceptions of the past is that Butler's initiative seems to have almost completely eclipsed recognition of the Lamarckian element in Spencer's thought. For most commentators today, Lamarckism is seen as an expression of an outdated vitalism. Perhaps in the non-Darwinian world, people would have retained a sense of the theory's more complex history and implications.

Parallelism and Orthogenesis

The American school of neo-Lamarckism to which Cope belonged reveals a very different manifestation of the theory in science. Despite its origin as a means of explaining adaptation, Lamarckism became associated with the idea of predetermined non-adaptive evolution known as orthogenesis. This model is very much a product of the formalist tradition that had flourished earlier in the century in Germany, a tradition imported into America by the émigré Swiss naturalist Louis Agassiz. Although Agassiz himself was an arch-creationist, the fact that his disciples could move rapidly into a form of evolutionism shows how easily the idea that nature is based on orderly patterns could be translated into a theory of predetermined evolution. Not surprisingly, the American followers of Agassiz developed a model of evolution very similar to that flourishing among the non-Darwinian German biologists.

To understand how Lamarckism could be turned into a theory of predetermined development, we must go back to the hypothetical starting point from which a line of specialization such as that leading to the giraffe's neck began. Initially, animals were free to choose their new habit, but once that commitment was made, their descendants were anything but free to innovate—they became locked into a behavior pattern they could not modify without harming themselves. They were on an evolutionary treadmill and could not step off, so the trend continued inexorably toward the maximum viable level of specialization. At this point the species was trapped by its extreme commitment to a particular way of life, and if the environment changed in a way that threatened that lifestyle, the species was doomed to extinction. For neo-Lamarckian palaeontologists such as Cope and the invertebrate palaeontologist Alpheus Hyatt, the theory explained what they saw as the rigid parallel trends displayed by many groups in the fossil record, trends so predetermined that they could lead to the appearance of more or less identical species on independent lines of development. This version of neo-Lamarckism challenged the concepts of divergent evolution and common descent. Similarity of structure might not be inherited from a common ancestor but could instead emerge when two parallel lines subject to the same trend independently reached the same stage in a preordained hierarchy of development.[21] Palaeontologists, including Cope's disciple Henry Fairfield Osborn, were still advocating parallelism in the early twentieth century.

Lamarckism also undermined the assumption that evolution was driven primarily by the relationship between organisms and their environment. Hyatt incorporated neo-Lamarckism into a vision of evolution driven largely by predetermined patterns of development. Trends that began as adaptive specializations gained a kind of momentum that drove a species beyond the point of maximum viable specialization into realms where its characteristics became overgrown and harmful. This was the theory of "racial senility" based on an extreme version of the idea that there is a link between the life cycle of an individual and that of its species.[22] Although it could be linked with Darwinism, the recapitulation theory was far more amenable to an explanation in Lamarckian terms. For recapitulation to work properly, the old adult stage must be preserved in the embryological development of

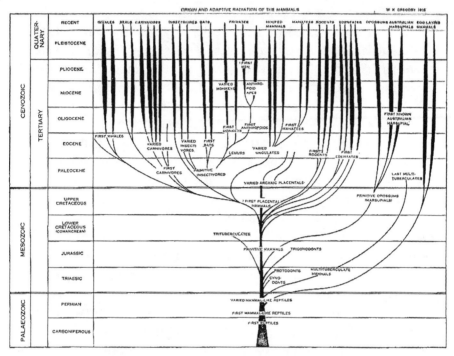

FIG. 114. ANCESTRAL TREE OF THE MAMMALS.

Figure 8 · Tree of mammalian evolution from Henry Fairfield Osborn, *The Origin and Evolution of Life*. Here Osborn shows a sudden radiation at the start of the class's history, after which each family evolves rigidly without further diversification as though driven by an internally programmed trend. He applied the same model of parallel evolution within each family.

the new species, and this would only occur if the new characteristic were added onto the end of the developmental process. This would be the case if a characteristic acquired by the adult was transmitted to its offspring by being added on as the last stage in the embryo's development. If variation is a distortion of the developmental process (as it would be if it were random), the later stages would be eliminated, leaving no clue as to ancestry. The recapitulation theory thus became associated more with Lamarckism than with Darwinism. The related idea that species have a youth when they are vigorous and adaptive, a maturity when they are a dominant life form, and an old age when they decline toward extinction, is a classic expression of the developmental viewpoint.

An oft-quoted example of racial senility was the so-called Irish elk, whose antlers were supposed to have become so big they eventually contributed to its extinction. Darwinians have a potential explanation of such structures through the theory of sexual selection. Apparently harmful characteristics such as the peacock's tail or the Irish elk's giant antlers might be useful in the competition for mates. Without Darwin, there would be no theory of sexual selection to offer a possible explanation of such structures, and even in our world, few naturalists took the idea seriously until the twentieth century. The theory of overdevelopment implies that although species become adapted to a particular way of life, their dependence on the environment is not so complete as to prevent the appearance of maladaptive extensions of a once useful trend—the resulting structures can reach substantial levels of overdevelopment before extinction occurs. Such a position would have been unthinkable to Darwin, and it is likely that even in a non-Darwinian world, naturalistic thinkers such as Spencer and Huxley would have regarded it as highly implausible.

The figures we associate with Darwinism would not, however, reject the whole concept of predetermined variation. Huxley assumed that there would be laws governing variation—the production of new characteristics would not be completely undirected (or "random," in the terminology of anti-Darwinists), as Darwin seems to have believed. Huxley was prepared to believe that there were only certain channels along which evolution could be directed, although any channel that led toward positively non-adaptive consequences would not get very far. His disciple E. Ray Lankester viewed the nature of variation this way in 1896: "It is a most misleading and erroneous view which has somehow got its head up, that any given animal varies in all its parts in every conceivable way, so that any result may be produced by selection. On the contrary, variation in every group is characteristic, is limited by the already selected and emphasized characteristics of the group. Every part therefore of an animal which varies, varies not 'equally around a norm,' but varies in accordance with the constitutional tendency of the organism, which may be called its ancestral bias or group bias."[23] Lankester had by now abandoned his early support for Lamarckism, but he was clearly unwilling to give up the idea of directed evolution. If a biologist with a reputation as a Darwinist could write of variation in these terms, we can understand how the idea of directed evolution gained a hold on the imagination of a whole generation of biologists. How much stronger this hold would be in a world where natural selection had never been debated and there was no outcry against Darwin's idea of "random" variation. In such a worldview, the possibility that built-in variation trends might eventually push a species toward overdevelopment and eventual extinction would seem even more plausible.

Huxley and Lankester were Darwinians because they acknowledged that the environment imposed constraints on which variation trends would succeed. For the extreme anti-adaptationist, there were no such constraints—the whole idea of a struggle for existence that might prevent inbuilt variations from gaining a place in the world was false. The theory of orthogenesis emerged as an alternative not just to Darwinism but to any theoretical perspective supposing that all evolution must be adaptive. "Orthogenesis" means evolution directed in a straight line, and the presumption of its most enthusiastic proponents was that those lines of development had no relevance to the organisms' adaptive needs. If there were adaptive modifications, Lamarckism could explain them, but for most evolution, adaptation was simply irrelevant. The German biologist Theodor Eimer popularized the term "orthogenesis" in opposition to Darwinism and tried to demonstrate the existence of such trends in the coloration of lizards and butterflies. Eimer even explained the phenomenon of mimicry as the outcome of two unrelated insect species independently becoming subject to the same variation trend. It had nothing to do with one gaining an advantage by piggybacking on the warning colors of another inedible form, as Bates and Darwin supposed. Like the American neo-Lamarckians, Eimer rejected Darwinism outright. In a world without Darwinism, it is reasonable to suppose that the evidence apparently implying directed evolution would have been explored with even greater enthusiasm and would have provided a more substantial counterweight to the adaptationist approach.

By its very nature, the theory of inbuilt variation trends focused on non-adaptive evolution and parallelism. Eimer's claim that mimicry was the product of parallel trends in two insect species was a small-scale example of the idea that led Mivart to suggest that different types of mammals could have evolved independently from separate reptilian ancestors. In a world without Darwin, such ideas would seem far more reasonable, since the rival notion of common descent would have lacked one of its greatest sources of inspiration. This point has important consequences for our assessment of evolutionism's cultural impact in the alternative universe. When applied to the origins of humans, the theory of parallelism could be used to suggest that the similarities between the races were independently evolved, allowing the races to be described as distinct species that did not share a recent common origin. This position was developed in the late nineteenth century and was still being promoted in the 1920s by the American palaeontologist Henry Fairfield Osborn, one of the last great defenders of orthogenesis. In the non-Darwinian world, parallelism might have played an even more prominent role in the wider debates over human origins.

Saltationism

The assumption that evolution could generate characteristics that were adaptively neutral or even harmful was also shared by another group of naturalists who rejected the principle of continuity. Darwin followed Lyell's insistence that all changes must be slow and gradual, and in principle both Lamarckism and orthogenesis were consistent with this position. But for many naturalists, the assumption seemed unwarranted. We know that variation does occasionally occur discontinuously, as in the case of monstrosities, and it seemed possible that such drastic changes might occasionally create new species. This was the theory of evolution by leaps, or saltations. Even Huxley thought Darwin overstated the case for believing that "nature makes no leaps," and throughout the late nineteenth century, many biologists thought it more probable that evolution proceeded by jumps. Saltationism had the advantage that it preserved the reality of species—there was no implied blurring of the boundaries as there is in a theory of gradual transformation. Some Lamarckians accepted an element of discontinuity. Cope, for instance, argued that the pressure to change might at first be resisted and then, after many generations, give way suddenly to allow the accumulated effects to be transferred into a species' inheritance. Orthogenesis might result from a series of saltations, each transforming development in a consistent direction. Most saltationists refused to believe that the struggle for existence was powerful enough to suppress new characteristics even if they had no adaptive value.

In a recent study of the debate over the origin of language, historian Greg Radick has pointed out the influence of saltationism on late nineteenth-century evolutionary thought. The Darwinian assumption that human speech must have emerged gradually was widely rejected in favor of the claim that it was the product of a sudden transformation of the brain and nervous system. Radick argues that we have consistently tended to ignore the extent to which Darwinian gradualism failed to take hold, and it seems reasonable to suppose that in a world without Darwin, the saltationist position would have enjoyed even more support.[24]

In our own world, opposition to the principle of continuity was particularly strong among those biologists who tried to understand the nature of heredity. Francis Galton, for all that he stressed the power of selection acting on normal variation to "improve the breed," nevertheless insisted that a saltation was needed to form a new species. He compared the species to a polygon balanced on one face—normal variation is represented by rocking about on that face, but real evolution requires a disturbance strong enough to topple the polygon onto a new face.[25] Several of the biologists associated with the emergence of modern genetics approached the topic though the assumption that new characteristics were created discontinuously. Hugo De Vries, William Bateson, and Thomas Hunt Morgan all became saltationists, and Morgan wrote a provocative anti-Darwinian text, his *Evolution and Adaptation* of 1903.

The theory implied that new species established by saltations could perpetuate themselves whether or not they were better adapted to the environment than the parent form. Like orthogenesis, saltationism necessitated the rejection of adaptation as the driving force of evolution. Bateson and Morgan both attacked the idea that the struggle for existence could prevent a mutated form from establishing itself—if enough mutated individuals were born, they would constitute a successful breeding population distinct from the parent form. Of the three, only De Vries accepted that the less well-adapted mutations would eventually be eliminated, and that only after a number of generations.[26] For Bateson and Morgan, the whole adaptationist paradigm was misconceived—the environment simply didn't have the power to suppress the production of new characteristics by mutation.

The link between saltationism and Mendelism must be taken into account as we turn to consider how the theory of natural selection might eventually have been discovered in a world without Darwin. In our world, the reemergence of Darwinism after its eclipse was made possible by the creation of a new theory of heredity—genetics—that was fundamentally hostile to Lamarckism. This hostility was a product of internecine warfare between the two main alternatives to natural selection, Lamarckism and saltationism. If this tension was endemic to the non-Darwinian paradigm, we can presume that it would emerge in a world without Darwin—although perhaps more slowly. Eventually, though, there would be a saltationist challenge to Lamarckism, and this would create a new climate of opinion eventually opening a way for the discovery of natural selection.

6

WHENCE NATURAL SELECTION?

If non-Darwinian evolutionism flourished in the late nineteenth century, when would the theory of natural selection have been introduced? Two conditions would have to be satisfied to place scientists in a position where they would have had to postulate such a mechanism. First, the developmental perspective's tendency to see evolution as the unfolding of built-in trends would have to be minimized in favor of the rival approach focused on dispersal and local adaptation. Second, doubts would have to emerge about the only other conceivable hypothesis of adaptive evolution, the Lamarckian inheritance of acquired characteristics. Both of those requirements would be satisfied by developments taking place in the decades around 1900. That is exactly what happened in our world, and it led to the reemergence of Darwinism from its eclipse—although "eclipse" is a misleading term given that the theory of natural selection had never enjoyed much success among life scientists. The traditional areas of science devoted to reconstructing the history of life on earth now witnessed major changes of emphasis and important new discoveries that made it increasingly difficult to portray evolution as a collection of rigidly programmed trends. Even more significant for the overall history of science, new disciplines emerged, including a new science of heredity—genetics—that challenged the Lamarckian paradigm.

There are good reasons for supposing that both of these developments would have occurred whether or not the Darwinian selection theory was in play. In a world without Darwin, circumstances favorable to the discovery and promotion of that theory would have emerged, just as they did in our world. But instead of encouraging a revival of interest in natural selection, they would have allowed space for its actual discovery. Sometime in the 1890s or the early decades of the twentieth century, the idea of natural selection would be proposed and taken up by naturalists who were becoming aware of the weakness of the theories they had adopted for so long. The selection theory would emerge at a point coinciding with its reinvigoration in our own world, thanks to the synthesis with genetics. But in the non-Darwinian world, that theory would be a new one, not a revival—and it would not, of course, be called "Darwinism."

There would already be a limited model of selection available in the non-Darwinian world—the belief that weaker species and varieties are eliminated when better-adapted forms invade their territory. We know that such a weeding-out process was accepted even by naturalists who openly rejected any suggestion that natural selection acting on individual variations could provide a positive force for generating new varieties and species. Everyone was aware that human interference with the natural world was beginning to cause the extinction of species, either directly or indirectly through the introduction of alien species. The dodo of Mauritius and the great auk (a large seabird) were classic examples. The Cambridge professor of zoology Alfred Newton was particularly concerned about this phenomenon and campaigned for the conservation of threatened species.[1] People also recognized that even without human action, natural processes occasionally allowed species to cross geographical barriers and invade new territories, often with disastrous consequences for the existing inhabitants. This was a standard theme in biogeography, and this model of selection would be readily available in the world without Darwin. But how easy would it be to transfer this model of competition between species to the level of individuals within a population? If Darwin's insight really was original, it may have been more difficult than we might imagine.

Alfred Russel Wallace had conceived a theory of natural selection, but it focused mainly on competition between varieties, not between individual variations. In a world without Darwin, he would have promoted the idea of selection among geographical varieties as an explanation for adaptive evolution and might have made a brief foray into the area of individual variation. However, my suspicion is that this would not have been a major part of his project, and any theory he promoted would have lacked several of the key themes that made Darwinism so effective. Wallace would not have used the analogy between natural and artificial selection, and his religious views prevented him from developing any real sense of the cruelty and indifference of nature. Even if he suggested the theory of natural selection acting at the individual level, it would not have had the influence it acquired through the *Origin of Species.*

At best, the idea would have had only a limited application, perhaps in those areas where Lamarckism ran into difficulties. One such area was the study of coloration in insects, a topic addressed by Wallace and in Henry Walter Bates's study of mimicry. Since insects cannot deliberately change their own colors, simple use-inheritance does not explain the adaptive coloration that many species show. But the effects might be explained by invoking selection among varieties rather than individuals, and this was especially likely since Wallace was working on the relationship between species and varieties at the time. Later in the century, E. B. Poulton at Oxford studied the same effects and became a staunch Darwinist precisely because the selection theory works so well in this area. But in a world without Darwin, the topic of insect coloration might have attracted less attention, and we cannot be sure that an Oxford professor would have been working on it at this later stage, let alone that he would have recognized selection acting within populations. To explain how the idea of selection could emerge as a key player in the game, we must imagine a sequence of events that would focus the attention of a wide range of biologists on the need to find an alternative to Lamarckism.

What other factors might encourage biologists to think along lines similar to those followed by Darwin? Who would link topics as far apart as biogeography and animal breeding? Based on what we know about the individuals who made major contributions in our world, I suggest we should have real doubts about their ability to create a selection theory without Darwin's influence. Only toward the end of the nineteenth century did all the necessary details start to come together, and then only from a variety of different sources.

The factors leading to this transition were both scientific and nonscientific. In part they lay in new discoveries by palaeontologists and geologists that focused more attention on the correlation between major transformations in the earth's environment and equally dramatic episodes of extinction and rapid evolution. The sheer weight of fossil evidence was also undermining the neat linear trends constructed by the advocates of orthogenesis. These discoveries led to the decline of the various non-Darwinian positions in the late nineteenth century, allowing the adaptationist perspective to fight back against developmentalism and creating a climate of opinion receptive to new initiatives for explaining how adaptive evolution might work.

A more fundamental transition within the life sciences is associated with what has been called the "revolt against morphology" at the end of the century. At this time, life scientists dismissed as unscientific the techniques that had underpinned phylogenetic research and the recapitulation theory, and their emphasis shifted to the experimental study of the actual processes of variation, heredity and evolution. These new studies linked up with a growing sense that heredity was of crucial importance. They undermined Lamarckism by demanding hard experimental evidence for "soft" heredity and created a climate in which it was essential to provide an alternative explanation of adaptive evolution.

Forces external to science also promoted the new model of heredity. An equivalent to the Darwinian theory would not have been conceived until someone recognized an analogy between the artificial selection practiced by animal breeders and what happened when a natural population faced an environmental challenge (this is why Wallace doesn't fit the bill). That analogy would become apparent only when the ideology of hereditary determinism began to raise its head in the late nineteenth century. As a consequence of broad social pressures (too broad to be seen as mere by-products of scientific thinking), the middle classes became dissatisfied with the Lamarckian assumption that individuals can improve themselves and hence contribute to the progress of the race. The idea that heredity rigidly predetermines character came to the fore, giving rise to a proposal that the state should control the breeding of the human population to prevent the proliferation of the "unfit"—the movement known as eugenics.

It was no coincidence that the concept of heredity in biology was clarified only at this time, so similar moves would almost certainly have been afoot even without the impetus of Darwinism. Whether the scientists' response would have produced the exact equivalent of what we call genetics is uncertain, but there would certainly be theories based on "hard" heredity (the realization that the transmission of characteristics cannot be modified by an organism's environment or upbringing). These theories would develop in a context in which the practice of artificial selection in both animals and humans was seen as the only way of improving the stock. Lamarckism would come under greater experimental scrutiny and would be found wanting—so what would take its place as a theory of natural, adaptive evolution? Surely at this point someone would have recognized that there could be a natural form of selection producing effects similar to those attributed to Lamarckism but in a manner consistent with the new hereditarianism.

In our world, the emergence of genetics led to a conflict between the proponents of the new theory of heredity and the comparatively small number of biologists still defending the Darwinian selection theory. Only in the 1920s and 1930s was that conflict resolved through the creation of the genetical theory of natural selection. In the counterfactual world without Darwin, we need to look for an equivalent process, which, in this case, leads to the actual discovery of the selection theory. If the revolt against morphology also led to a program of studying variation and its inheritance in wild populations, I think that someone would have spotted the potential analogy with artificial selection as invoked by the eugenics movement. The result would be a theory of natural selection—but it would be a theory that would have emerged in a form much less likely to lead to a confrontation with the other non-morphological research programs.

A crucial difference between our own world and the counterfactual one would modify how the theory of natural selection would be perceived. The history of genetics as we know it includes an episode in the early twentieth century when, in America especially, the new science turned its back on the study of embryology, paying no attention to how genes actually produce the characteristics for which they code in the developing organism. Many biologists now believe that this lack of coordination between genetics and development produced an impoverished version of the selection theory, a situation only recently corrected by the emergence of evolutionary developmental biology (evo-devo). In the non-Darwinian world, it would be less easy for the developmental aspects of genetics to be marginalized. Evo-devo would have been part of the synthesis from the start, a situation that did, in fact, obtain in Germany and France.

This reversal of the sequence of discovery would have had major implications for how the new theory was perceived. In our world, Darwin forced everyone to confront the most materialistic form of evolutionism before they had even come to terms with the general prospect of a natural development of life on earth. Darwinism was identified as a theory of "random" changes winnowed by a brutal struggle for existence. But if natural selection was discovered later, after everyone had become accustomed to the general idea of evolution, it might have seemed less threatening. The new theory would have been incorporated into a synthesis that retained the more fruitful aspects of the developmental viewpoint. It would have been obvious from the start that the effects of genes were mediated by a developmental process that was to some extent purposeful, so there would be less temptation to see natural selection as a process that blindly weeded out mutations produced by chance.

The new theory would also have had a less visible public profile because it would almost certainly be pieced together by a number of biologists, with each step being openly discussed in the scientific literature. There would be no blockbuster impact similar to the one sparked by the *Origin of Species* and no single individual who could be pushed forward as the figurehead of a revolution. There would be time for everyone to come to terms with the new idea, and since it would not be presented as a leap into materialism, its implications would seem less threatening. The synthesis of natural selection and evo-devo would appear as a product of the regular development of science, not as a bombshell threatening to undermine the worldview to which everyone was accustomed.

NEW DIRECTIONS

New initiatives in the life sciences would almost certainly create a general climate of opinion less favorable to the anti-adaptationist aspects of the developmental approach. The non-Darwinian world would have witnessed at least as great an emphasis on phylogenetic research as ours did, and the increasingly sophisticated work of palaeontologists and geologists would have transformed the way evolutionists visualized the history of life on earth. Just as in our own world, evolutionary morphology would be eclipsed as a technique for uncovering relationships because more direct evidence was becoming available from the fossil record. This would take place at exactly the same time as most laboratory biologists would be turning against morphology and looking for new methods of studying development, heredity, and variation through direct observation and experiment. But these new methods would not automatically generate insights compatible with the palaeontologists' discoveries—indeed, the study of fossils would be itself dismissed as unreliable by many of the new disciplines. As in our own world, there would be conflicts between the various scientific disciplines and no guarantee that new theories of heredity would be presented in a form compatible with an adaptationist approach to evolution.

The New History of Life

In the 1870s and 1880s, the fossil evidence favorable to evolution increased dramatically as geologists explored more of the earth's surface. The opening up of the American West was just one example of a new territory yielding fossils that would transform the argument, as Huxley pointed out in his discussion of O. C. Marsh's sequence demonstrating the evolution of the horse family. At first, many of the palaeontologists who described these fossils saw them as evidence for non-Darwinian versions of evolutionism based on predetermined trends and parallelism. Marsh's great rival in the race to exploit the fossil beds of the west, Edward Drinker Cope, was a leading exponent of neo-Lamarckism and orthogenesis. Many argued that the fossil record supported evolution, but not the Darwinian model of branching, adaptive change.

As more evidence accumulated, though, it became clear that the trends were not as direct as they had initially seemed. When there were only a few fossils available for a group, it was easy to imagine them arranged in a straight line leading toward a final form. This is how they were often arranged in museum displays or textbook diagrams. But as more fossils were unearthed, they seldom fit into the neat patterns that neo-Lamarckians hypothesized. The sheer weight of evidence began to favor a more haphazard, unpredictable model of evolution that we associate with Darwinism. Something like the modern horse, for instance, seems to have evolved on a number of occasions whenever an open-plains environment has appeared, only to go extinct when the conditions have changed.[2]

It wasn't just the fossils that forced a change of worldview—geologists were become more experienced at working out the conditions under which rocks were formed. By reconstructing past environments, they could correlate evolutionary events with climatic changes. Lyellian uniformitarianism (on which Darwin relied) had to be modified to include the notion of dramatic and sometimes quite sudden climatic transitions driven by occasional bouts of intense mountain building. This was not exactly a revival of catastrophism, but there was a reaction against the idea that geological change has always occurred at the same rate. The concept of what we now call a mass extinction reemerged, and it was clear that in the uncrowded world following a climatic trauma, the evolution of the survivors exploded into a diversity of forms. The classic example was the final extinction of the dinosaurs followed by the flowering of the mammals. The mammals had actually appeared much earlier but had never had a chance to expand their range until the dominant reptiles went extinct, clear evidence of the unpredictability of progress.

This was not how Huxley and Haeckel's generation had imagined evolution—they had seen development as a much more continuous process. I like to think that Darwin himself would have appreciated the new evidence had he survived to witness the discoveries. Certainly the rather simple developmental vision of the mid-nineteenth century did not derive primarily from the influence of his theory. It was the product of a worldview obsessed with the progressive unfolding of history through predictable trends. By the early twentieth century, a much more "modern" vision of the history of life on earth had emerged, one that supported a vision of evolution driven by unpredictable environmental stresses rather than inbuilt trends. In our world, this new model of life's history helped to prepare the way for the revitalization of Darwin's theory in the middle decades of the twentieth century. In a world without Darwin, the developmental model might have been more influential at first, but eventually it would have succumbed to the ever-increasing weight of fossil evidence in favor of unpredictability.

We can judge the impact of fossil discoveries from the fact that Henry Fairfield Osborn, one of the leading supporters of developmental trends, pioneered the concept of adaptive radiation to explain the sudden diversification of the mammals after the extinction of the dinosaurs. He played a leading role in the transformation of the earth sciences that allowed for closer correlation between fossil species and their environments. Yet he remained wedded to the view that once the diversity of forms was established, their subsequent development was predetermined. His commitment to a developmental worldview did not prevent him from seeing at least some of the evidence for massive traumas in the history of life.[3] Osborn's position gives us a good reason to presume that in a world without Darwin, the implications of the new discoveries would not have gone unrecognized.

The rival biogeographical approach focused on migration and adaptation. It would certainly be available in the world without Darwin, because it could exploit the simplest form of Lamarckism as an explanation of local adaptation. In the counterfactual world, this position might have enjoyed less success, because Darwin had provided one of its most charismatic sources of support. But Wallace and others would still have ensured that this view of the history of life was alive as an alternative to developmentalism. By the early decades of the twentieth century, it was becoming more difficult for the last defenders of orthogenesis such as Osborn to claim that the record showed clear evidence of linear trends. The biogeographical approach gained in plausibility as more fossil evidence from around the world was correlated with the growing geological evidence for major disruptions in the earth's physical environment. The history of life was episodic and irregular, just as the biogeographers predicted. These developments would have occurred as long as there was a continued expansion of geology and palaeontology, and we can thus be confident that in the world without Darwin, the early twentieth century would have been ripe for the emergence of a new explanation of adaptive evolution.

The Revolt against Morphology

By itself, a new picture of the history of life was unlikely to generate a theory of natural selection. Even in our world, many naturalists and palaeontologists were so committed to Lamarckism that they refused to acknowledge new initiatives in other areas of biology. The most important pressures leading to a reconsideration of the evolutionary mechanism would come from the emerging crisis in the study of heredity. To understand how this crisis threatened the plausibility of Lamarckism, we have to look to wider trends in the development of the biological sciences, including what historian Garland Allen has called the "revolt against morphology" that began in the last decades of the nineteenth century.[4]

In the heyday of morphology, scientists used comparative anatomy and embryology to reconstruct the relationships between living things and infer their evolutionary history. Morphologists were not experimentalists, and their descriptive approach did not encourage detailed investigation of how evolution works on a day-to-day basis. Within this framework, theories such as the inheritance of acquired characteristics could flourish without being required to produce experimental evidence—indirect support was enough. Lamarckism got an easy ride in the 1860s and 1870s even in our world, where the main source of detailed argument about heredity was the effort to discredit Darwinism. Without that alternative to rock the boat, the scientists would have accepted the Lamarckian model of "soft" heredity (inheritance that can be modified by the parents' own experiences) virtually without question.

Eventually another factor began to throw light on the question of heredity: a growing dissatisfaction with the whole morphological approach within which developmental evolutionism had flourished. In part this was a reaction to the failure of morphological techniques to provide unambiguous answers to key phylogenetic questions, such as the origin of the vertebrates. In the 1890s, William Bateson abandoned his research on this topic in disgust at the failure of morphologists' efforts to decide which similarities were signs of ancestral relationship and which were mere parallelisms. Bateson turned to studies of animal and plant breeding to throw light on the actual processes of variation and heredity, leading him ultimately toward genetics. Others with similar dissatisfactions turned to the statistical analysis of variation in wild populations, or to the experimental study of the cellular processes responsible for reproduction. They dismissed morphology as the mere description of dead organisms. A real science of life, they now claimed, would use experiment and mathematical analysis to reveal the actual processes at work in variation, heredity, and evolution.

The rejection of evolutionary morphology undermined the credibility of the recapitulation theory, which had become one of the main lines of indirect evidence for Lamarckism. It also called into question the casual assumption that changes in the adult organism must somehow affect the mechanism by which characteristics are passed on to the next generation. Once biologists started to look carefully, it turned out to be far more difficult than anyone had suspected to provide hard evidence that the Lamarckian effect actually worked. In the early twentieth century, Bateson led the geneticists' campaign to discredit the Lamarckian experiments of Paul Kammerer in what has become known as the "case of the midwife toad."[5] The scientific community subsequently dismissed Kammerer's efforts to demonstrate a Lamarckian effect in salamanders and the midwife toad as inadequate or even fraudulent. In Britain and America, geneticists nailed their colors to the mast of the new concept of "hard" heredity—the claim that the material responsible for transmitting characteristics from parent to offspring cannot be modified by changes to the parents' bodies. In the process, some of them lost interest in embryology altogether. It was enough to know that characteristics were transmitted unchanged from one generation to the next; there was no need to know how those characteristics were produced in the developing organism. Transmission genetics was all the rage, while what was later called developmental genetics—how the genetic information was actually unpacked to form a new organism—was ignored. This oversimplified model of heredity was subsequently transferred to the emerging Darwinian synthesis, and it has only lately been undermined by the creation of evolutionary developmental biology, or evo-devo.

Historical studies of the origins of genetics have suggested that this impoverished model was not an inevitable stage in the process, but was instead a reflection of the highly specialized nature of the American scientific community. In Germany, France, and, to a lesser extent, Britain, the attitudes we now associate with evo-devo remained influential and scientists maintained an interest in the implications of development for evolution.[6] This opens up the possibility that in a world without Darwin, a much later discovery of the selection theory might have been blended more seamlessly into the surviving components of the old developmental approach. We can now appreciate that the developmental perspective was not all bad. Although it encouraged some naturalists to minimize the role of adaptation and search for purely internal causes of evolution, modern science has had to rediscover, at the cost of some controversy, the importance of individual development for a full understanding of how evolution works. In a world where the selection theory was not introduced until several decades after 1859, these controversies might have been avoided and we would see natural selection as an extension of the older program, not as a totally revolutionary insight.

CLARIFYING HEREDITY

Curiously, Bateson began his campaign to study variation and heredity by presenting it as a rejection of Darwinism. By "Darwinism," he meant the casual assumption that all characteristics must be adaptive and also the erection of what he deemed to be flimsy hypotheses about evolutionary relationships that were beyond experimental proof.[7] But in the end, his program turned out to be far more destructive of Lamarckism than it was of Darwinian natural selection, and after some initial confusion, genetics became the salvation of the selection theory. Could a parallel process lead to the actual discovery of the selection theory in a world where it had not been introduced previously? This seems plausible, because several areas of investigation were pointing in the right direction. Francis Galton and his followers in the eugenics movement applied the concept of hard heredity to the study of human populations. The same idea emerged simultaneously in the work of biologists, most notably August Weismann, trying to understand the cellular basis of reproduction. Somewhat later there was a flurry of interest in breeding experiments as a means of tracing inheritance, which in our world let to the emergence of genetics. We need to uncover what each of these three innovations contributed to the new model of hard heredity, and then ask whether they were likely to generate the idea of natural selection in a world where it was not already available. In fact, it seems unlikely that the selection theory could have emerged directly from the study of heredity, because in our world the revival of Darwinism also required an input from naturalists investigating variation in the field, not in the laboratory or breeding station. The new model of heredity was essential if Lamarckism was to be replaced by natural selection, but on its own, that model was not enough to prompt the necessary connections.

Hard and Soft Heredity

In our world, two initiatives began to focus attention on heredity as a determinant of character during the 1880s. The work of Francis Galton in Britain and August Weismann in Germany led to the emergence of what we now call the concept of "hard" heredity, which challenged the casual assumption that the Lamarckian effect must be real. These developments were clearly associated with Darwinism, so it is not immediately obvious that they would have occurred in a world without Darwin. But other factors, both scientific and ideological, were also involved, and there are good reasons for supposing that these would have been sufficient to prompt similar innovations even without the selection theory. The transformation in scientists' thinking about heredity would be crucial for the development of evolutionism, but it was much more broadly based and would probably have occurred in a very similar manner in the counterfactual world.

It is hard for us to realize that the concept of heredity in the modern sense emerged only in the nineteenth century. Everyone knew that the basic characteristics defining species and varieties or races were preserved through the generations. But the concept of individual characteristics being transmitted from parent to offspring was recognized only in extreme cases of family peculiarities. The one that attracted the most attention was hereditary insanity, widely feared and concealed. Of course, the animal breeders knew that heredity extended far more widely, but few naturalists saw their work as a model (which is why Darwin's interest in this area was so unusual). As far as most people were concerned, individual differences stemmed not from heredity but from upbringing and the environment, from "nurture" rather than from "nature," to use the now popular terminology. The liberal social philosophy of Herbert Spencer depended on the view that individuals could improve themselves through effort and education—they were not rigidly limited by the capacities they had inherited. Spencer and the Lamarckians also assumed that such self-improvements could be passed on to future generations. This was "soft" heredity, and it was only in the late nineteenth century that people both in science and in society began to question it. Francis Galton's contribution was to argue that the possibility of self-improvement was extremely limited. All of a person's physical and mental characteristics were predetermined by what they had inherited from previous generations via their parents. The Lamarckian effect was discredited almost as an afterthought once scientists accepted that there were in fact no significant acquired characteristics.[8]

Galton's inspiration came from his commitment to a wider social movement that led the middle classes to doubt the effectiveness of reform as a means of improving the human situation. As an explorer in Africa, Galton was struck by the apparently rigid differences between the races. When his early career in geography and exploration collapsed, he looked for another area in which to make his mark as a scientist and settled on the study of heredity within family groups. He became convinced that individual character was rigidly determined by heredity. Galton's book *Hereditary Genius* of 1869 insisted that brilliant people inherited their qualities from their parents; the fact that they probably also received a better education was irrelevant. He went on to develop sophisticated ways of studying the variation of characteristics within the human race, pioneering the statistical approach to the gathering and analysis of information about variation and heredity within the population. He also promoted a social philosophy known as eugenics, basically applying the principle of artificial selection used by animal breeders to the improvement of the human race.

Galton highlights two factors crucial for the formulation of the theory of natural selection: hard heredity and the analogy with artificial selection. In a world without Darwin, would he have developed the theory of natural selection himself in the course of the 1870s? Galton was Darwin's cousin, and he quite deliberately constructed his project as a way of making his reputation among men of science such as Huxley. It's conceivable that had his cousin not been there to serve as an inspiration, he might have chosen another topic entirely. But this is unlikely given his early interest in racial differences and his subsequent obsession with statistics. Thanks to Darwin, he was alert from the start to the analogy between artificial and natural selection, and this would not have been available without the *Origin of Species*. But once he conceived the project to improve the human race by encouraging the "best" people to have more children (and the worst to have less), the model provided by the animal breeders would have been impossible to miss. Without necessarily digging into the detailed literature that fascinated Darwin, he would have begun to recognize the need for selection.

Artificial selection would thus enter the debate about evolution for the first time (since no one else would have promoted it in Darwin's absence). But would this have been enough to inspire Galton himself to discover and promote the principle of natural selection? The main stumbling block here is that even with Darwin's theory before him, we know that he did not believe that selection could transform a species into something else. Selection applied to normal individual differences could improve the breed within fixed limits, but like many of Darwin's critics, Galton thought it was incapable of altering the fundamental character of a species. He insisted that a saltation—a jump to something completely new—was needed to establish a new species. Given this aspect of his thought, it's hard to see him paying much attention to the possibility of selection operating in wild nature unless he had his cousin's theory to follow. Any hints developed by Wallace would have been ignored, because Wallace was emerging as a champion of social equality. The transition to a theory of natural selection would only become possible when those scientists not still trapped by the idea of a reality of species decided to take up Galton's statistical techniques.

Despite his clarification of the idea of hard heredity, the detailed theory of inheritance that Galton conceived was in no way an anticipation of genetics. His "law of ancestral inheritance" did not involve the transmission of hereditary unit characteristics from parent to offspring. Instead, he proposed that the organism received one half of its inheritance from its parents, one quarter from its grandparents, and so on in diminishing proportions from more distant generations. When Galton's follower Karl Pearson eventually sought to clarify the effects of selection, he developed a model of continuous change that led him to minimize the significance of new discoveries by the geneticists. Their theory of discrete unit characters was dismissed as being based on artificially produced breeds rather than on natural variation. The result was a controversy that delayed the synthesis of genetics and Darwinism. We cannot assume that in a world without Darwin, Galton would have been led to discover either Mendel's laws or the principle of natural selection.

Galton began to gather masses of statistical evidence for the variation of characteristics in the human population because the social program he proposed in the late 1860s was largely ignored by his contemporaries. They were unwilling to accept that all the efforts of individuals to improve themselves were useless; the assumption that people could improve themselves was crucial both to Spencer's ideology of self-help and to those who campaigned for social reform. Galton worked tirelessly to accumulate evidence for the determining role of heredity in human character. Eventually he was rewarded when a change of public attitude in the last decade of the century led to growing enthusiasm for the hereditarian viewpoint. Spencer's extreme laissez-faire individualism was now out of fashion, and the efforts of reformers seemed to be having little effect. Galton's eugenics movement at last gained momentum, which it and retained well into the twentieth century. These social developments were very broadly based and did not depend on new ideas in biology, so we can be pretty certain that they would have occurred even in a world without Darwin. But they did focus scientists' attention on the problem of heredity, allowing theories such as Galton's to be taken seriously.

The Germ Plasm

Changing social attitudes also focused attention on investigations of the cellular basis of heredity. We associate this development with August Weismann, who proposed the concept of the "germ plasm" in the 1890s to explain how hereditary characteristics are transmitted from a parent to a fertilized ovum. The conceptual foundations of Weismann's idea seem so obvious today that it is difficult to appreciate just how revolutionary his idea was. Even Darwin, who certainly saw the need for characteristics to be transmitted intact from parent to offspring, did not visualize heredity in the way we do today. It seemed obvious to Darwin and almost everyone else that the parent organism must actually manufacture the material substance from which the embryo of its offspring would develop. Darwin proposed a theory of "pangenesis" in which each part of the body buds off tiny particles that pass to the reproductive organs, where they are assembled to form the ovum or sperm. To describe the model in today's terminology, it was as though we each manufacture the genes we pass on to our children. The fact that we might transmit a family characteristic from previous generations would mean only that—because our body reflects that characteristic—we were producing genes that followed roughly the same pattern. But if we modified our body by adopting new habits (remember the weightlifter's bulging muscles), those modifications would also be passed on. This is why Darwin along with virtually everyone else at the time accepted the Lamarckian effect as real.

Weismann eventually rejected Lamarckism and proposed a new model of heredity implying that the effect simply could not work. He argued that the material substance that transmits the information needed to shape the embryo of an offspring is isolated from the rest of the body. He correctly predicted that this substance, which he called the "germ plasm," was located in the chromosomes of the cell nucleus. But he insisted that there was an impassable barrier between the germ plasm and the soma, the rest of the body. Whatever happened to the soma in the course of a parent's life, the germ plasm would remain unaffected, and any acquired character could not be transmitted. In effect, Weismann came close to the modern concept of genetic material (which we now know as DNA) being transmitted unchanged through the generations. We inherit genes from our parents, but they have been merely transmitted to us through them from previous generations. Each generation's bodies are made from the information stored in the germ plasm, but that information is passed on through an unbroken chain in the sex cells alone. This was a startlingly new concept of heredity, and it is small wonder that at first many found it so counterintuitive that they refused to believe that the body could be powerless to influence the hereditary material it would transmit.

Weismann pioneered an important aspect of the modern view of heredity, and to do it he had to accept natural selection as the only viable explanation of how adaptive evolution can work. By denying the credibility of Lamarckism, he was forced to become an arch-selectionist, and this tightly focused version of the theory became known as "neo-Darwinism." (Darwin himself was not regarded as a neo-Darwinist because he still accepted an element of Lamarckism.) If we try to imagine the sequence of events in a world where there was no selection theory available to Weismann, we have to ask whether he could have come up with the idea, and if not, whether he could have conceived his new model of heredity. I think there are serious doubts on the first of these points. If there was no well-articulated theory of natural selection thanks to Darwin, it is by no means clear that Weismann would have had the conceptual resources to create it for himself. This in turn means that the concept of the isolated germ plasms would be far more difficult for the biologists of the non-Darwinian world to develop. There would still be a general focus on the power of heredity thanks to the social developments, so the idea of an all-powerful substance responsible for hereditary transmission would almost certainly have been articulated eventually, but it would have been a more gradual process possibly spreading into the early twentieth century.

One reason for questioning Weismann's ability to step into Darwin's shoes as the originator (as opposed to merely the champion) of the selection theory is historian Frederick Churchill's observation that he was very much a thinker within the developmental tradition.[9] His early studies in the 1870s were intended to throw light on the mechanisms by which variant characteristics within species were expressed. He worked on color variation in butterflies and caterpillars, trying to show that the forces producing new patterns were not teleological—they were not inbuilt trends anticipating the organisms' adaptive needs. At this point, Weismann was concerned only to limit the power of the Lamarckian effect, not to eliminate it altogether. He accepted that changes in the environment might stimulate

186 · CHAPTER SIX

variation, but saw most of that variation as non-adaptive. Only after failing eyesight forced him to abandon his microscopic investigations did he turn to theoretical modeling of the process of heredity, articulating the concept of the germ plasm and insisting on the "all-sufficiency of natural selection" for adaptive evolution. Even then his complex views on the actual nature of the germ plasm—which in no way anticipated the modern notion of the unit gene—allowed for non-adaptive variations arising from competition between germinal elements.

Weismann was primarily an embryologist, and although he had extensive field collections of the species he studied, he seems to have had little interest in the phenomena of geographical distribution and speciation. Nor is there any evidence that he was prepared to study the work of animal breeders for clues as to how variation and heredity work. Two of the key resources that shaped Darwin's thinking as he moved toward the selection theory were thus unavailable to Weismann. In these circumstances, it is hard to imagine him putting forward a comprehensive theory of natural selection on his own, although conceivably he might have formulated a rudimentary version of the theory to explain the phenomena in which he was interested. Even if we assume that Wallace had already attempted to float a theory of individual selection, this would have been within a biogeographical perspective that was alien to Weismann. In the world without Darwin, there would have been at best only disjointed efforts to explore the possibility that natural selection was the driving force of adaptive modification in populations. There would be no coherent effort strong enough to make the theory a major player in the 1870s and 1880s.

The idea that an organism's development was predetermined by the information coded in its germ plasm would also have taken longer to formulate, although, given the attractiveness of this idea to the supporters of eugenics, it is hard to imagine that it would not have arisen at some point in the decades around 1900. As developments in cytology (the study of cells) pushed ahead, there was increasing focus on the chromosomes of the cell nucleus and their ability to control the development of the embryo. The theory of the germ plasm evolved into the ideology of hereditary predetermination—the organism was nothing more than an expression of controlling factors ("coded information," in modern terms) imprinted onto the chemical nature of the chromosomes. Some biologists saw this as a revival of the old theory of preformation, in which the new organism merely expands from a miniature enclosed in the fertilized ovum. Switch attention from the whole ovum to the nucleus, they argued, and imagine preformation in metaphorical rather than literal terms, and you have the core insight of the new approach. To the more enthusiastic advocates of this position— although not to Weismann himself—the question of how the information stored in the nucleus was developed into the body of the embryo could be largely ignored. This total repudiation of the developmental approach would eventually be transformed by new insights into how characteristics were stored and transmitted across generations.

Genetics

In our world, these insights were codified by the new science of genetics. The "rediscovery" of Gregor Mendel's laws of inheritance in 1900 focused attention on characteristics that are transmitted as fixed units from one generation to another. The new field, known first as "Mendelism," but soon renamed "genetics," represented another manifestation of the revolt against morphology. It was based on the experimental breeding of artificial varieties of plants and animals, a practice in which the distinct nature of hereditary characteristics is particularly obvious. In Mendel's own experiments, the colors of his peas could be yellow or green, but never a yellowish-green blend. His posthumous followers believed that their artificially controlled breeds held the key to understanding the whole process of heredity. They assumed that where the distinct nature of the genetic units is not apparent, the characteristics are being modified by transient and hence insignificant environmental factors. The model of nuclear preformation also had many followers. Once T. H. Morgan and his school had shown that genetic units were located on particular segments of chromosomes, it seemed obvious that the chemical nature of a unit somehow encoded information specifying the characteristic that must develop from it. Nuclear preformation was the rule, the complexities of the developmental process could be ignored, and—almost as an afterthought—the Lamarckian process became implausible because environmental modifications of development were assumed to be insignificant.

Traditional histories of Darwinism, including those published for the centenary of the *Origin of Species* in 1959, held that the discovery of genetics was the salvation of the theory of natural selection.[10] Genetics not only destroyed the credibility of Lamarckism, it also got rid of the old notion of "blending heredity," according to which the offspring is a smooth mixture of the characteristics of its two parents. If Darwin's original formulation of natural selection had a weakness, it was this model of heredity, as represented by his own theory of pangenesis. Critics such as Fleeming Jenkin showed that if blending heredity is correct, a favorable new characteristic will be diluted by half each time the individual possessing it reproduces, and it will soon be reduced to insignificance. Even if selection is boosting the reproduction of favored individuals, its effect steadily diminishes as advantageous characteristics are diluted. Better-adapted characteristics produced by variation are soon swamped by being blended into the mass of normal individuals making up the whole population.

The old story of Darwinism's revival held that genetics solved this problem at a stroke: characteristics do not blend, they breed true and reappear as whole units in all succeeding generations. Thus selection can transform a population by boosting the reproduction of individuals carrying favorable genes and eliminating those with maladaptive characteristics. New characteristics can be introduced from time to time by genetic mutations, accidental "copying errors" occasionally introduced into the system—the true source of the random variation postulated by Darwin. Even though many of these will be maladaptive, natural selection will hold them in check, while seizing on the rare mutations that carry some slight advantage.

This interpretation of the relationship between genetics and Darwinism presumes that if the theory of natural selection were not available in the late nineteenth century, it would certainly be discovered as soon as genetics revealed the true nature of heredity. Lamarckism would have to be abandoned, and it would be obvious that a process of differential reproduction operating among genetic units would be able to shift the relationship between genes in favor of those conferring reproductive success. The breeding programs of the geneticists would immediately suggest the model of artificial selection, allowing a natural equivalent to be conceived as an explanation of adaptive evolution.

As far as timing goes, this scenario is probably a good model for what would occur in the world without Darwin, but the process would be a complex one—as indeed it was in our world. The problem with the old history was that heredity wasn't the only or even the main reason why scientists rejected natural selection. The whole developmental tradition stood opposed to any theory depicting evolution solely as the product of local adaptation. Genetics emerged out of the non-Darwinian theory of evolution by saltations or sudden jumps, and most early geneticists did not believe that adaptation was a crucial factor in evolution. They certainly saw the relevance of artificial selection, but they had no reason to look for a natural equivalent. The necessary refocusing had to come from a different tradition that studied natural rather than artificial variation and sought to explain change in whole populations rather than in isolated artificial breeds.

More seriously, the old story of Darwinism's reemergence took for granted a version of genetics modeled on a beanbag in which differently colored beans are collected as isolated units. Selection merely alters the number of beans being transferred from one generation to the next, boosting the number of some colors and discarding others. This oversimplified model of genetics and evolution has been modified through the emergence of evolutionary developmental biology, which has reminded us that genes are not simple blueprints for individual characters. There is a whole sequence of developmental process by which genes interact with one another and feed into a cascade of transitions of immense complexity in order to produce a viable developing embryo.[11]

We now realize that those complex developmental processes are important for evolution after all. They may not include a role for simple Lamarckism, but they certainly entail factors that the model of nuclear preformation ignored. The developmental tradition would have been far more influential in the world without Darwin, which opens up the important prospect that in such a world, the development of genetics and the discovery of natural selection would occur in ways that did not entail a temporary marginalization of development. We have a perfect model for this situation in what actually happened to genetics in most countries except America. Our histories of genetics have been written from an American perspective, and as Continental historians have been telling us for years, that perspective ignores what happened in their countries. In France and Germany especially, genetics and the selection theory did not achieve the same level of dominance, and the developmental tradition was never completely eclipsed. If the oversimplifications of the beanbag model could be avoided, evo-devo would be there from the start, and natural selection would have had to fit into the framework it defined.[12]

To imagine how this might have come about, let's start with an obvious source of counterfactual possibilities: the breeding experiments that Gregor Mendel performed in the 1860s to establish his laws governing the transmission of unit characteristics from one generation to the next. Would Mendel have discovered his laws in a world without Darwin, and if not, would this have delayed the subsequent emergence of genetics at the end of the century? This nexus of alternatives may not be quite as crucial as one would expect, because recent histories of genetics have challenged many of the old assumptions about Mendel's role. He certainly did not conceive of the gene as a material unit, and it is by no means clear that he saw his laws as a comprehensive new model of heredity. The "rediscoverers" of his laws in 1900, Carl Correns, Hugo De Vries, and Erich von Tschermak, read a good deal of their own conceptualization of heredity back into Mendel's original papers. This allowed them to hail him as the founder of the new science, perhaps to head off an otherwise damaging priority dispute. In the words of historian Robert Olby, Mendel was no Mendelian, and it seems reasonable to suppose that the new studies around 1900 would have generated the laws sooner or later without his influence. Another reassessment of Mendel suggests that his real motive for undertaking his experiments was not to understand heredity but to investigate the possibility that new species might be produced by hybridization. He was driven to this by distrust of Darwinism, raising a very real possibility that in a world without Darwin, he would have had no reason to perform his breeding studies at all.[13]

Even so, with or without Mendel, there are good reasons for supposing that the revolt against morphology would have led some biologists to look toward experimental breeding as a source of information about heredity and variation. Given the increased fascination with the model of hard heredity at the turn of the century, such a program would generate new theories to explain what scientists observed. Would the result be similar to what we call genetics, including equivalents to Mendel's laws detailing how unit characteristics are transmitted?

Historian Greg Radick raises the interesting counterfactual possibility that under quite plausible scenarios, alternative theories could have emerged and possibly triumphed. Francis Galton and his follower Karl Pearson were working on models of heredity that did not involve anything like the concept of unit characteristics. Pearson's collaborator W. F. R. Weldon was developing a theory of his own that would similarly have marginalized the effects studied by the geneticists. Pearson and Weldon were openly hostile to William Bateson, the biologist who translated Mendel's papers into English and coined the term "genetics." But Weldon died young at a crucial phase in the debate, and Radick argues that this unexpected event left the way open for Bateson and his followers to triumph, establishing genetics as the new orthodoxy. The counterfactual possibility is that in a world where Weldon lived to fight on, genetics would not have had so easy a ride and other models focused on different effects less easily explained in terms of unit characteristics would have become established instead.[14]

Radick's suggestion plays on a recognition that personalities do matter and that the elimination of a key participant in a debate might allow a different outcome. The early theories of the geneticists were grossly oversimplified and worked only for carefully selected and highly artificial experimental subjects. Weldon was coming from a very different direction that in our world led a small number of naturalists to stay loyal to Darwinism. He would probably have begun his studies of wild populations even in a world without Darwin, and this very different observational foundation would have supported alternative models of heredity. But I find it hard to believe that there would have been no equivalent to genetics, even if its hegemony had been challenged by alternative explanations. Experimental breeding programs will almost certainly focus on clearly visible character differences, and most of the founders of genetics came from a background in the saltation theory encouraging them to believe that new characteristics appeared as discrete units. From the conviction that characteristics arise by jumps, it is a small step to believing that they must be inherited as discrete units.

This was in fact the route by which several figures involved in the foundation of genetics came into the program. Hugo De Vries had made his name by propounding his "mutation theory" in which new varieties and species appear suddenly when their original form gives birth to significant numbers of individuals displaying a new characteristic. William Bateson had turned his back on phylogenetic work to write a book in 1894, *Materials for the Study of Variation*, which focused on characteristics that appeared as discrete units. His examples included new varieties of flowers with a different number of petals, which almost certainly could not appear by the slow expansion of a new petal over a series of generations. The leader of the American school of genetics, Thomas Hunt Morgan, passed through a phase in which he supported the mutation theory and—like Bateson—derided the idea that adaptation played a key role in evolution. Given that the role played by saltationism would have been even greater in a world without Darwinism, it seems inevitable that experimental studies of heredity would have led to the idea of the unit characteristic and then to an elucidation of something like Mendel's laws. If Bateson was one of the leaders of this movement, the new theory might even be called "genetics."

It is the next step in the process that leaves open the best possibilities for a counterfactual world where heredity theory and evolutionism were deflected onto a new track. In our world, genetics in Britain and especially in America became openly hostile to Lamarckism. They focused attention solely on the transmission of unit characteristics from one generation to the next, but the whole question of how the genes actually generated their characteristics in the developing embryo was set aside as too complex to be tackled immediately. Development was regarded as largely irrelevant to the most productive research program. If environmental factors could modify the genes' effects slightly, that was of no permanent significance either for breeding or for evolution. This is why Bateson led the campaign to discredit the Lamarckian experiments of Paul Kammerer in the case of the midwife toad.

The strongest endorsement of nuclear preformationism came when T. H. Morgan, a belated convert to Mendelism, founded an experimental program designed to show that the laws of inheritance could be correlated with the behavior of the chromosomes in the reproductive cells. In effect, a gene could be localized as an element along a string making up the length of a chromosome. There was, as yet, no inkling of how a gene actually coded the information for the characteristic it represented, let alone how that information fed into the development of an embryo, but for Morgan and his school, these issues were irrelevant. They could make progress by adopting a deliberately oversimplified approach focused on the one problem that could be tackled with existing techniques. Genes could be located on the chromosomes, and this information could be used to explain the effects observed in breeding experiments. Transmission genetics could be established as a coherent discipline without worrying about the complexities of how genes influenced development.

It was this highly specialized and in some respects oversimplified approach to genetics that, in our world, was incorporated into the synthesis of the new science with the theory of natural selection. But in a world where the selection mechanism was only just being recognized, it might have been much less easy for the ideology of nuclear preformation to gain so dominant a hold. In his *Styles of Scientific Thought*, Jonathan Harwood argues that what English-language historians have tended to regard as the "normal" pattern by which genetics developed was in fact a highly localized research program centered in America and to a lesser extent in Britain. He contrasts the ways in which genetics was consolidated in Germany and America, showing that the specialized techniques of the Morgan school worked well in an expanding country where practical applications were important and new departments and laboratories easy to found. In Germany the academic community was not expanding, and many scientists held to an ideal that valued breadth both in scientific vision and in wider cultural interests. Here the oversimplified approach of the Morgan school was dismissed precisely because it failed to address the broader question of how genes affect the developing organism.

In such a climate, it was possible for theories at variance with nuclear preformationism to thrive. Many geneticists believed that the cytoplasm surrounding a cell nucleus also played a role in shaping development. Hostility to Lamarckism was by no means as profound, and even those who rejected the inheritance of acquired characteristics were prepared to accept that the complexity of development might allow some environmental influence. In Britain there was also less enthusiasm for the Morgan school, Bateson in particular despising the chromosome theory as too materialistic. Here too some geneticists retained an interest in problems of development and in processes that sidestepped the notion of nuclear preformation. Conrad Hal Waddington's later theory of genetic assimilation is an example of a theory that allowed a role for individual adaptation without invoking actual Lamarckism.

When these points are taken into account, the simplified model of the Morgan school begins to look like an exception, formulated in the unique academic environment of early twentieth-century America, rather than the only route by which genetics could advance. The new science could and did become established without going down the route of absolute nuclear preformationism, and in a world where the developmental tradition was even more strongly established than in our own, that route would be the more obvious path of scientific advance. In the non-Darwinian world, it would have been much less easy for anything like the Morgan school to gain the dominance it did, even in America. Whatever the temptations of overspecialization, they would have been less obvious in a world where the Lamarckian theory had been completely dominant up to that point.

The Great War severed relationships between the German and English-speaking scientific communities and generated suspicions that took several years to dissipate after the armistice. This breakdown allowed the American school to ignore issues that were still taken seriously by many German scientists. Would there have been a Great War in a world without Darwin? Modern critics of evolutionism see the war as a manifestation of social Darwinism, but it's hard to imagine that a scientific theory could be the principal driving force. The political and economic rivalries that led to the war were too profound to be defused by the mere lack of a scientific metaphor. So there would still have been a war and a temporary breakdown of communication between German and English-speaking geneticists. My argument is, however, that even in America, the developmental view of evolution would have been harder to eliminate in the non-Darwinian world, making it much less likely that a period of isolation would have led to such an overspecialized program of research on heredity. The two communities would have reunited much more smoothly after hostilities ceased.[15]

None of this is meant to imply that genetics would have coexisted with simple Lamarckism. The geneticists would naturally have been inclined to see the gene as something that could not respond directly to an organism's efforts and new habits. Bateson and others might well have organized a campaign to undermine the credibility of experimental efforts to demonstrate a direct Lamarckian effect, and this would have prompted naturalists to look for an alternative mechanism of adaptation. But most biologists would not have tolerated a blanket rejection of all processes in which development might play a role in shaping the outcome of genes' influence. In a world where the developmental tradition was more dominant, it would have been unthinkable to abandon that tradition completely—it would have been obvious to all that this was throwing the baby out with the bathwater. The ideas we now identify with evolutionary developmental biology would have retained a hold even among English-speaking geneticists and would have been part of the cultural environment within which the theory of natural selection would at last emerge.

NATURAL SELECTION AT LAST

Genetics (or its equivalent) would cast doubts on the credibility of the direct Lamarckian effect, but would not by itself generate a theory of natural selection. New ideas about heredity emerged in part out of an enthusiasm for the concept of evolution by jumps or saltations, reflecting the antiadaptationist position characteristic of the developmental model of evolution. Because their experiments were conducted under highly artificial conditions, geneticists were reluctant to believe that the pressure of the environment could impose limitations on which new characteristics could thrive in the wild. Even though they were well aware of the power of artificial selection, there was little incentive for them to develop a modification of this idea to explain adaptive evolution. The same would likely be true in the world without Darwin. The focus on adaptation, and hence on the need to find an alternative to simple Lamarckism, would have to come from another branch of biology more in tune with fieldwork and the study of populations living in a natural environment. In our own world, there was such a program, yet another manifestation of the revolt against morphology, which turned to the detailed investigation of variation in wild populations rather than artificial breeds. It provided the main source of support for the beleaguered Darwinian theory of natural selection. In a world where there had been no Darwin to launch that theory in the 1860s, it would be these biologists who would be the first to come to a real appreciation of what the selection model had to offer.

Biometry versus Genetics

When William Bateson abandoned evolutionary morphology in the 1890s, his first attempt to study the actual processes of variation and heredity involved a fieldwork project in central Asia, where he tried to demonstrate the inheritance of environmentally stimulated variation, a form of Lamarckism.[16] The results were largely negative, so he turned instead to the study of artificial breeds. He also abandoned the view that adaptation is crucial for evolution, and he developed the distrust of Lamarckism that later motivated him to campaign against Kammerer's experiments with the midwife toad. His friend W. F. R. Weldon made a similar transition from morphology to fieldwork, but his observations were more productive and led to a new approach to the analysis of variation known as biometry. But Weldon did not reject the utilitarian paradigm, although he too became suspicious of a Lamarckian interpretation of how populations adapt to changes in their environment.

In our world, Weldon became one of the few champions of the theory of natural selection during the 1890s, leading to a break with Bateson and hostile relations that persisted until Weldon's untimely death in 1906. His studies of variation in populations of snails and crabs sampled huge numbers of individuals, revealing a significant range of variation in each wild population (something that is essential for natural selection to work, provided the variations are inherited). Weldon was in fact generating the now familiar bell curve revealing a concentration of individuals around the mean value for a continuously varying character (such as height in the human population) and smaller proportions tailing off to the extreme on either side. In the case of the crabs he studied in Plymouth harbor, there was a systematic modification of the range of variation over a number of generations, apparently linked to changes in the environment caused by artificial dredging. Weldon was convinced that what he was seeing was the effect of natural selection shifting the range of randomly distributed variation, just as Darwin had predicted.

His conviction only grew stronger as he began to collaborate with the statistician Karl Pearson, who was already applying his mathematical talents to the analysis of Francis Galton's data on human variation. Sampling large populations, as Weldon was doing, required a more sophisticated form of statistics, which Pearson supplied. Pearson in turn subscribed to Galton's ideology of hard heredity and became an enthusiastic supporter of eugenics. But Pearson soon abandoned Galton's assumption that his "law of ancestral heredity" implied variation about a fixed mean, with saltations being required to create a new mean and hence a new species. He became a convert to the theory of natural selection and showed that the theory was perfectly adequate to explain adaptive evolution on the basis of Galton's non-Mendelian theory of heredity. When the geneticists began to argue that all significant variation is discontinuous, Pearson was able to show that the approach he and Weldon were developing could accommodate evidence for limited discontinuity, while insisting that most natural variation follows the bell-curve model. But the hostility between Weldon and

Bateson dragged Pearson into a bitter dispute that polarized biometry and genetics into mutually incompatible programs.

How would this situation pan out in a world where there was no well-developed theory of natural selection for Weldon to draw upon? I suggest that here at last we have a situation in which the mechanism of natural selection would be fully recognized and exploited. The collaboration between Weldon and Pearson was an obvious move for both of them because each had something from which the other could benefit—a huge selection of data on Weldon's side and the relevant statistical skills on Pearson's. Given Pearson's commitment to Galton's hereditarianism and the concomitant awareness of the role of artificial selection in the eugenics program, we have a situation in which it would immediately be obvious that there might be a natural process of selection acting on the variation that Weldon had revealed. Here, if anywhere, is where natural selection *ought* to have been either discovered or exploited properly for the first time. Perhaps there would be some precursor elements derived from Wallace and Weismann, but biometry brought together twin insights derived from artificial selection and the study of populations adapting to new conditions in the wild, just when Lamarckism was looking increasingly vulnerable. Natural selection would emerge as the only conceivable alternative to explain adaptive evolution.

By saying that natural selection "ought" to have been discovered or exploited at this point, I am implying that in our world the "natural" process of scientific development was somehow distorted by the unique and unpredictable input from Darwin. No one else was in a position to duplicate Darwin's achievements in the earlier part of the century, and most of his contemporaries found the theory either hard to understand or totally unacceptable. It is not unreasonable to imagine that in a world without Darwin, it would not be until the 1890s that all of the pieces would fall into place, with the selection theory appearing as the obvious solution to the pressing need for a better explanation of adaptive evolution. In putting together the study of geographical distribution (what we now call ecology) and animal breeding, Darwin pulled off a coup that no one else at the time was in a position to do. In effect he jumped the gun, combining the necessary ingredients several decades before general developments within biology made this combination obvious to everyone else. The result was a kick start to the emergence of scientific evolutionism, but also a debate that poisoned the atmosphere by forcing everyone to confront a materialistic theory that few felt comfortable with. In the non-Darwinian world, that debate would not have happened, and evolutionism would have developed a little more slowly and with a more developmental emphasis, but in the end the conditions ripe for the discovery of natural selection would have emerged.

A Synthesis—but Not as We Know It?

Only in the 1920s were significant efforts made to heal the rift between biometry and Mendelism. Several biologists including R. A. Fisher, J. B. S. Haldane, and Sewall Wright began to realize that unit genes could explain continuous variation if several of them influenced the same characteristic. Their effects would interact in a large population to form the bell curve observed for variation in many natural populations. Natural selection could then work by boosting the frequency of those genes that conferred a characteristic tending toward the adaptive end of the frequency range while whittling down the proportion of those responsible for a maladaptive influence. The result was Fisher's classic book of 1930, *The Genetical Theory of Natural Selection*. For practical reasons, both Fisher and Haldane treated genes as independent units circulating independently within a population, new ones being created every so often by mutation. By this time, however, it was already becoming apparent that genes did not act in isolation, and the formulation of the new theory worked out in America by Sewall Wright took this effect into account.[17]

The final stage in the emergence of the modern Darwinian synthesis came when fieldworkers and palaeontologists began to use the new version of natural selection to throw light on the problems of biogeography, speciation, and evolutionary trends. The Russian émigré Theodosius Dobzhansky translated the abstract mathematics of the genetical selection theory into propositions that fieldworkers could comprehend and use. Naturalists such as Ernst Mayr, a refugee from Germany in the United States, realized that the Lamarckism they had formerly taken for granted as the explanation of adaptive evolution was no longer plausible, and he took up the selection theory instead. In Britain, Julian Huxley, T. H. Huxley's grandson, promoted a similar integration of the new techniques and traditional fieldwork, giving the new approach its popular title in his 1942 book, *Evolution:*

The Modern Synthesis. This was a synthesis between genetics and Darwinism but also between laboratory science and fieldwork.

Some of the original founders of the synthesis were willing to allow a limited role for non-Darwinian effects, but soon there was a "hardening" of the selectionist position leading to a paradigm that permitted only the natural selection of independently operating genetic units.[18] All of the effects once associated with the old developmental tradition were suspect, including saltations, constraints on variation, and any form of environmental influence on heredity. The new Darwinism did indeed seem to promote an image of evolution as a completely trial-and-error process in which the environment sifted copying errors (mutations) in order to allow the very few that conferred an advantage in reproduction to increase their proportion within the population and thus modify the species in an adaptive direction.

Would there be a parallel process in the world without Darwin? Once the selection theory had emerged within the kind of detailed fieldwork pioneered by Weldon, it seems inevitable that it would be linked to whatever developments were taking place in heredity theory. These would include at least some recognition of the role of genetic units, even if not in the very rigid form postulated by the Morgan school. And a genetical theory of natural selection would immediately prove attractive to fieldworkers and other biologists who were becoming aware of the weaknesses of simple Lamarckism and the anti-adaptationist excesses of the developmental tradition. It seems likely that many of the figures with whom we are familiar in our own history would play a role, perhaps with German-language scientists being more prominent.

The most obvious difference is that the modern evolutionary synthesis of the non-Darwinian world would not be able to trace itself back to a breakthrough made by a single iconic figure. Without Darwin there would be no Darwinism, and that means more than just the lack of a term—it means there would be no clear-cut foundation that could be identified in hindsight as defining the essence of the evolutionary paradigm. Evolutionism would have emerged gradually, its passage into science and modern culture eased by the developmental approach's focus on progress and other purposeful or orderly trends. Natural selection too would have appeared not as a bombshell challenging all traditional assumptions about the world, but as the product of normal scientific research programs jockeying for position and each contributing its bit to the overall process. Genetics and natural selection would be seen as having gradually transformed developmental evolutionism, purging it of its more teleological extremes and bringing it steadily closer to a naturalistic (but perhaps less explicitly materialistic) worldview. We do not identify our evolutionary synthesis with a single figure because we can see that many different biologists from many different backgrounds helped to create the synthesis. In the non-Darwinian world, that sense of a cumulative and cooperative endeavor would apply to the whole evolutionary movement, going right back to its origins. There would be much less sense of a revolution, least of all one owing its origins to a controversial event that happened back in the mid-nineteenth century. Evolutionism would be seen as the product of the regular processes of scientific discovery, of normal rather than revolutionary science (to use T. S. Kuhn's terms) or at least of a continuous series of mini-revolutions rather than one big one.

As a consequence of this more continuous emergence from the developmental tradition, both genetics and the evolutionary synthesis would be much less likely to pass through a phase in which the gene was regarded as little more than a blueprint for a single characteristic. In our world, genetics and neo-Darwinism did pass through such a phase, especially in America, and we have had to correct this oversimplified model of genetic activity through developments leading toward evo-devo. Neo-Darwinism at first focused on the relationship between a breeding population and the environment in which it lives, seeing organisms as mosaics of individual unit characteristics that can be added to by mutation or subtracted from by natural selection. But it has long been obvious that genes do not act in isolation. They interact with the environment and with each other, sometimes in unpredictable ways that generate apparently new characteristics. Ernst Mayr emphasized exactly this to undermine the logic of the original "beanbag" model of the organism as an assemblage of discrete genetic units. Even so, many popular accounts of the theory still use the beanbag model for the sake of simplicity, perpetuating the misconception that there is a single gene for each identifiable character.

Evolutionary developmental biology offers a new perspective differing significantly from that of traditional neo-Darwinism, and even now the two are not completely integrated. In the 1980s we began to recognize that the expression of genes can only be understood through a better understanding of how they control—and are controlled by—the developmental process that generates a new organism. The evo-devo or epigenetic approach is focused on the developmental processes that maintain the basic integrity of the major animal groups, and it sees evolution in terms of modifications of those processes. These developmental processes may owe their origin to the genes, but their basic structures were assembled in the distant past and they still determine the ways in which modifications can be made. Some systems control the development of the same organ even in widely differing groups, for example the homeobox or *Hox* genes that control leg or eye formation in creatures as different as humans and fruitflies. Such a combination of deep conservatism with striking diversity of application was certainly not anticipated by the geneticists who saw the genome as a collection of independently modifiable unit characteristics.[19]

The predetermined developmental pathways impose limits on what kind of modifications can be successfully introduced by mutations. Thus variation is not always random but may in fact be constrained to move only in certain directions. Such constraints may be a far cry from the rigid trends of the old theory of orthogenesis, but there is an element of continuity linking the new ideas back to the old developmental viewpoint. Enthusiasts for evo-devo argue that it may sometimes be possible for relatively small genetic changes to trigger a cascade of developmental modifications producing a significantly new outcome, a concept reminiscent of the old theory of saltations. The differentiation of the major animal phyla, for instance, may not have occurred through the summing up of many small, adaptive changes, but through a burst of radical restructuring triggered by such switches. The emerging consensus suggests that efforts to reintroduce the more extreme anti-Darwinian aspects of the old developmental program will not succeed. But the parallels between the old and the new paradigms are still important and reinforce Ron Amundson's point about the need to see continuity in their histories.[20]

Epigeneticists also recognize nonnuclear components to heredity and complex processes that allow an organism to respond to the environment in ways that can be passed on to future generations. Thus something like the old Lamarckism may be possible, and some scientists have suggested that Kammerer's discredited experiments with the midwife toad may have actually detected such epigenetic modifications.[21] In a host of different ways, evo-devo has reawakened interest in topics that were once central to the nineteenth-century developmental model of evolution, although only its more enthusiastic advocates expect it to overturn Darwinism's emphasis on the need for all variations to be tested by the environment.

Now consider what might have happened in a world where the developmental approach had become even more strongly entrenched in late nineteenth-century evolutionism. The anti-adaptationist aspects of this approach would still have come under fire toward the end of the century as palaeontologists and field naturalists exploited the biogeographical model of evolution. Increased interest in the power of heredity would have generated theories similar to genetics, undermining the credibility of the simple Lamarckian mechanism. But as these forces created a climate of opinion in which the theory of natural selection could emerge, the additional momentum gained by the developmental tradition would make it much less easy for the investigation of heredity to become so specialized that it could focus solely on the transmission of unit characteristics. Physiological and developmental genetics would remain part of the theoretical package, as indeed they did in the Germany with which we are familiar. The beanbag model of genes transmitted as totally independent units from one generation to the next would not have gained so strong a hold.

As a result, when the theory of natural selection emerged, it would do so in a climate of opinion favorable to the possibility that the input from mutations is mediated through processes that influence how new genes are expressed. It would also have been appreciated that nuclear genes are not the only features of interest—the cytoplasm surrounding the nucleus also plays a role and may be more susceptible to environmental influence. Even with selection imposing the demands of adaptation, evolution would have been understood as a complex interaction between genes, developmental pathways, and the environment. The kind of synthesis toward which modern biologists are moving as they come to terms with the impact of evo-devo would have been in place from the beginning. Natural selection would have emerged in a form that blended smoothly into the still valuable remnants of developmentalism. The seesaw of opinion that in our world led from developmentalism to neo-Darwinism and then back to evo-devo would have been avoided.

The non-Darwinian world would eventually achieve a synthesis similar to the one that our scientists are now creating, containing essentially the same components derived from epigenetics, the adaptationist approach, and the idea of natural selection. But those components would have been put together in a different sequence, with the emergence of the selectionist component being delayed until the early twentieth century and being incorporated into a still flourishing developmental tradition. Quite possibly a workable synthesis would emerge even more rapidly than in our world, even though its earliest phase would have been delayed. The absence of Darwin would delay the appearance of the selection theory, but its eventual emergence would be far less disruptive. Everyone would already be familiar with the general idea of evolution and would be used to thinking of it in positive terms made possible by the developmental model's imagery of purposeful and orderly progress. Far from being a great shock to the system, natural selection would be greeted as a useful solution to the problem generated by the increasingly obvious weaknesses of the more extreme neo-Lamarckian and orthogenetic aspects of developmentalism.

In the world without Darwin, the sort of synthesis we are only now beginning to realize might already have been in place by the mid-twentieth century. The great paradox implied by this counterfactual hypothesis is that although Darwin was amazingly prescient in his theorizing, proposing insights that would not be appreciated by most scientists for half a century or more, he threw the debate over evolutionism off course by introducing a concept with which most of his contemporaries could not cope. In popular parlance, Darwin was ahead of his time, and evolutionism might have developed more smoothly if biologists had been left to explore a less materialistic version of the theory as a stepping-stone to the more radical vision that would eventually have to emerge. No one during Darwin's time was prepared to deal with a theory so obviously at variance with traditional ideas about nature and its Creator, and Darwin's radical insight generated the pendulum swings of opinion that have bedeviled the history of evolutionism in our world. It seems counterintuitive, but we should consider that Darwin's explanation of evolution might have been better left for a later generation of scientists who would have been able to incorporate it more smoothly into their thinking. Perhaps great revolutions are not always the best way of achieving major breakthroughs, especially if they require the scientific community to grapple with too many radical ideas at once.

Darwin's theory frightened his contemporaries because it implied that the world was governed by chance and struggle. It became an integral part of the scientific naturalism that T. H. Huxley and his followers used to challenge traditional religious beliefs. Modern neo-Darwinism to some extent revived the image of natural selection as a game of genetic Russian roulette, with the organism living or dying according to the luck of the draw in the lottery of inheritance. But in a world where selection was introduced in full recognition of the fact that the effects of the genes are mediated by complex and essentially purposeful developmental processes, this image would be much less obvious.

Darwin drew upon the Malthusian image of a world ruled by scarcity and struggle to promote his theory. He certainly modified that image by making struggle a creative force, although he was by no means the only thinker to do this. But natural selection does not depend on the notion of nature as a harsh and relentless force for its credibility—even Wallace argued against the Darwinian image of nature as a scene of endless cruelty and suffering. The specter of Malthus had largely evaporated by the early twentieth century, and the founders of the new Darwinism were much less inclined to invoke the struggle for existence as the driving force of natural selection. As R. A. Fisher and the founders of population genetics showed, natural selection will work as long as there is a differential in the rate at which genes reproduce. The unfit have to reproduce less often, but they do not have to die (a point that Darwin also recognized in his theory of sexual selection).

In a world without Darwin, natural selection would develop in a context defined not by Malthus but by the newly emerging science of ecology.[22] Its focus would be on the interaction between populations and their complex environments, a point that Darwin himself recognized in his metaphor of the "tangled bank" of what we would call ecological relationships. In the non-Darwinian world, this would be the framework within which natural selection would be conceived from the start; Weldon's work, for instance, clearly focused on how a population relates to its environment. Conceived in this context, the implications of the selection mechanism would seem much less threatening to traditional moral values. Both within science and in society at large, the emergence of the selection theory would have been less disruptive because it would not have been associated with the darker aspects of early nineteenth-century thought that had triggered Darwin's unique insights.

7

EVOLUTION AND RELIGION

A Conflict Avoided?

Historians present the Darwinian revolution as a classic illustration of the fact that science is deeply entwined with wider cultural developments. Just as biologists found inspiration in models and analogies available through nonscientific sources, their ideas resonated beyond the sphere of biology. This is to say that scientific theories are not just abstract models of interest to researchers; they have implications that shape the attitudes of both scientists and nonscientists alike. The different conceptual traditions within which evolutionists formulated rival evolutionary theories have their own wider implications, as do the particular theories themselves. Most of these philosophical, religious, and ideological implications have already been identified, providing us with a basis on which to construct a counterfactual history of how a non-Darwinian evolutionism would have developed in areas outside of science. We begin with the religious debates.

Many theologians opposed evolutionism even before Darwin published, but the *Origin of Species* touched a particularly raw nerve, and Darwinism has been a target for much negative rhetoric ever since. There are many different levels at which the devout Christian[1] might object to the theory, and the counterfactual project requires us to be clear about which relate to Darwinism and which to the more general idea of evolution. Some issues seem to have become crucial only at certain points in time, again opening up counterfactual possibilities. The complaint that Darwinism (and indeed the whole complex of historical sciences) undermines the credibility of Genesis was not much in evidence among Darwin's early opponents. Many educated people at the time had already accepted that geology required some aspects of the sacred text to be interpreted allegorically.

Some early twentieth-century fundamentalists were not, in modern terminology, Young Earth Creationists. The claim that the earth is only a few thousand years old was not revived until the middle decades of the twentieth century. Here is one factor that we can write out of any counterfactual history of the early phases of the debate.

Even without an input from biblical literalism, however, it is evident that the general idea of evolution would arouse the antagonism of many believers. Some key issues would have to be addressed even if the theory of natural selection was not in play. These issues were already obvious in the opposition to Chambers's *Vestiges* in the 1840s. People didn't like being told they were descended from apes—the idea seems to have evoked an almost visceral distaste even among some who were not very religious. More seriously from the theological point of view, the claim that humans evolved from an animal ancestry raises serious problems about the status of the human soul. If the soul is a spiritual element defining the human personality, but the animals are "the brutes that perish," how can that new property have appeared in the world without divine creation? Any evolution theory implying continuity between our animal ancestors and the first humans raises this problem. Evolutionists often assumed that mental faculties gradually intensified over the course of animal evolution, while conceding that some new applications of those faculties emerged in the first humans. But the implication is still there: the human personality is the product of the physical organization of the brain and nervous system, which has become more complex over the course of evolution. Other areas of science were also raising this issue, especially those areas that seek to understand our mental life in terms of brain activity. In this regard, people attacked Chambers for supporting phrenology—an early form of neuroscience—as well as for his evolutionism.

Evolutionists and liberal religious believers have always found it easier to reconcile their positions when they assume that the process generating new forms of life is progressive and purposeful. The human mind (or soul) then appears as the high point of the overall trend and can be understood as its intended goal. This allows one to argue that the Creator established the natural laws governing evolution as an indirect means of producing humankind. In effect, the element of design or purpose implicit in the idea of miraculous creation is transferred into the laws of nature.

Darwinism presented those seeking such a compromise with a major problem. Natural selection doesn't look like the kind of process that a wise and benevolent God would use to achieve His ends. It starts with variation that is undirected (often caricatured as "random") and proceeds via ruthless struggle for existence to achieve nothing more than adaptation to the local environment. Admittedly, adaptation was all that Paley's utilitarian concept of design explained. But evolutionism needs an element of long-range progress if it is to fulfill an ultimate goal such as the emergence of humankind. Natural selection mimics Paley's limited concept of purposefulness without actually being purposeful, which is why it was so useful as a rhetorical device for Huxley and the scientific naturalists, and for their modern counterparts. But Huxley's campaign against organized religion was in part a move to establish professional scientists as a new source of authority, replacing the church.[2] He was on the lookout for any argument that would undermine the credibility of traditional religious beliefs, and natural selection was too good an opportunity to miss—even if he did not think it could be the main mechanism of evolution.

Any divergent, open-ended process threatens evolution's ability to achieve any long-term goal. The adaptationist approach to evolution encapsulates an element of unpredictability, whatever the process of adaptation at work, because it makes every step subject to the hazards of migration and environmental change. Natural selection merely highlights the problem by making variation itself haphazard. Naturally, those who hope to find purpose in evolution have always found the Lamarckian theory of the inheritance of acquired characteristics more congenial. Lamarckism is still a mechanism of adaptation, but variation is driven by the actual needs and activities of organisms and does not have to be winnowed out by struggle. This looks much more like the kind of process a benevolent God would have instituted. More generally, Lamarckism was much easier to associate with the alternative formalist tradition in biology, especially in regard to the recapitulation theory with its model of evolution as the unfolding of a pattern resembling the development of the embryo toward maturity. Here evolution becomes much more predictable and orderly—again, features that one might expect to be incorporated in God's plan for the history of life on earth.

The counterfactual approach has something valuable to offer here because we can ask how religious believers would have reacted if evolutionism had become dominant in science long before the theory of natural selection came into play. If Lamarckism and developmentalism played the main roles in helping to establish evolutionism during the 1860s, they would have posed much less of a threat to liberal Christianity. A move toward reconciliation had already begun in the previous decade, and if Darwin had not intervened, the slow conversion would have continued both in science and among the religious community. Darwin gave a powerful stimulus to the scientific debate, but he did so by proposing a materialistic version of evolutionism that renewed the suspicions of those who wanted to preserve the idea that a Creator had designed the universe. In the resulting debate, Darwinism became the bête noir of religious thinkers, a threat to everything they believed. Evolutionism only began to gain support among liberal Christians when they recognized that natural selection was not the only theory that scientists were exploring and that its alternatives were much more congenial to their perspective. Without input from the *Origin*, the process would have proceeded far more peacefully and a working compromise would have been in place before natural selection was discovered sometime around 1900. A selectionism formulated at this later date would also have seemed less threatening because it would have been incorporated into the developmental approach rather than displacing it.

With no Darwinism to serve as a bogeyman frightening religious thinkers, evolutionism might not have become quite so crucial an issue for American fundamentalists in the 1920s. After all, they could blame many other sources for the modernism they distrusted, both in science and elsewhere. Here the counterfactual projection becomes more speculative, but it seems not unreasonable to imagine a world where evolutionism and religion would be much less openly at war. Liberal Christians have no problem with the theory anyway, and their hand would be strengthened if the most potent symbol of the apparent threat was absent from the public perception. Conservatives would still be opposed, of course, and wider cultural movements would almost certainly ensure the rise of the Young Earth position in the later twentieth century. But its supporters might be focusing more on other topics, if only because without the specter of Darwinism, the other targets would seem more tempting. Perhaps geology would move to center stage in the resistance to modern science—if the Young Earth position is considered valid, evolutionism drops out of the equation almost as an afterthought.

Critics of Darwinism (and some of its more extreme supporters) like to present the situation in terms of black-and-white alternatives, or more precisely in terms of one-to-one relationships between scientific theories and their philosophical or ideological implications. For example, if you are a scientific Darwinist, you must also be an atheist and/or a social Darwinist. The relationships between theories and consequences are actually much more complex. We can interpret a theory in several different ways, emphasizing different aspects with contradictory implications. Similarly, the same religious or ideological position can often draw inspiration from different scientific theories. This complexity can play the devil with the counterfactualist enterprise, except when it is used deliberately to challenge simple-minded assumptions about the wider meaning of theories. We can imagine a different trajectory in science that has very little effect on wider debates because in our world, the proponents of various positions often simply use alternative theories to gain scientific credibility. This is exactly what we shall find in the case of social Darwinism in the next chapter.

In religion too there are multiple possible relationships. It is difficult but not impossible to reconcile Darwinism and Christianity.[3] Christians who believe the world to be suffering as a consequence of original sin have occasionally seen natural selection as just the sort of process that might work in such a world. The rival Lamarckian theory is widely interpreted as less materialistic because it seems to imply that evolution follows the purposefully chosen activities of the animals themselves. Yet Lamarckism was also an integral component of Herbert Spencer's naturalistic evolutionism, widely seen as a component (along with Darwinism) of the worldview against which later neo-Lamarckians were reacting. To complicate matters further, Spencer also managed to smuggle the struggle for existence into his Lamarckism, thereby checkmating any suggestion that it is a purely Darwinian idea. This having been said, the fact that most liberal religious thinkers found it difficult to take Darwinism on board does allow us to imagine a counterfactual universe in which the apparently less materialistic alternatives smoothed the way for acceptance of the general evolutionary position.

APES AND SOULS

It was inevitable that much of the evolution debate would be fought on a battlefield of human origins. "Is man an ape or an angel?" was Benjamin

Disraeli's famous question.[4] But Darwin's theory was not the only trigger for a reopening of this debate in the 1860s. New discoveries in archaeology forced everyone to confront the possibility that our distant ancestors were primitive "savages." Evolutionism merely extended the link further down the chain, from savage to ape-man to the apes and eventually by implication to the lowest form of life. What then became of the human soul? Was it really just a manifestation of physical processes going on in the brain? Evolutionists were not the only scientists who had to confront this issue.

The Ape Connection

The assumption that evolutionism entails the transformation from apes to humans predates Darwinism. Lamarck had implied that we had passed through an apelike stage in our evolution. Chambers did not stress this link in his *Vestiges*, but the idea had already caught the public imagination. Huxley and Owen were already arguing about the closeness of the relationship between humans and apes by the time Darwin published, and the debate would have rumbled on even without input from the *Origin*. With or without Darwinism, the idea of an ape ancestry would have become associated with evolutionism in the course of the 1860s. All too often, people assumed that the link was with the newly discovered gorilla, widely portrayed as the most ferocious of the apes. A host of cartoons from this period parody the ape-human relationship and have been widely reprinted in modern studies of the Darwinian revolution. Some of those cartoons would almost certainly have been published without the stimulus of the *Origin*. I have to concede, though, that some would not: the ease with which the elderly Darwin with his beard and bushy eyebrows could be caricatured as an ape helped to make him the figurehead of the evolutionary movement as well as to cement the ape link in the popular imagination.[5]

The link was a product of the evolutionary movement as a whole, however—it did not depend on Darwin's particular theory. The Darwinian model of divergent evolution makes it plain that we cannot have evolved from any of the living great apes. The fact that we are related to the apes means only that we have all diverged from a common ancestor: the apes are our cousins, not our parents. Admittedly, the human branch must have changed far more than any of the others, so the common ancestor (if its

fossils were discovered) would almost certainly be classified as an ape, although of more generalized structure than any of the living apes. This point
should have been obvious to all when the evolutionists began to disagree as
to which ape species was our closest relative. On anatomical grounds it was
far from obvious that it was the gorilla. Darwin and many others chose the
chimpanzee, while Haeckel popularized the view that we are closest to the
orangutan. This was why Eugene Dubois went to the Far East in the 1890s
to search for what had become known as the "missing link." Here he found
the remains of *Pithecanthropus erectus* (now known as *Homo erectus*), widely
thought to throw light on how the human branch had acquired its upright
posture and bigger brain.[6]

All too often, though, scientists tended to think in terms of a more or
less linear ascent from something like one of the great apes up to modern
humankind. This oversimplified image caught the public imagination, and
it encapsulates the most basic form of the developmental model of evolution. The old "chain of being" had arranged all the animal species in a
single hierarchy from amoeba to human, with the apes naturally fitting in
as the link immediately below ourselves (or, to use another analogy, the
next lowest rung on the ladder). Many popular diagrams purporting to illustrate the course of evolution used this linear model. There was sometimes a nod toward the image of a branching tree, but the tree always had a
central trunk running up through the apes and to the humans—everything
else was a side branch of lesser importance. Far from being a clear derivation of Darwin's theory, or indeed any theory of divergent evolution, this
was a continuation of an older tradition encapsulated in the developmental
model. One expression of this model was the recapitulation theory's analogy between human evolution and the development of the human embryo.
Here all the lower species were treated as steps on a single predetermined
upward path, and the apes would fit in as the last step before the human
form was achieved.

The developmental model thus highlighted one of the most damaging
aspects of evolutionism as far as many ordinary people were concerned: the
idea that we had emerged from something as disgusting or even frightening as an ape. For religious thinkers, this almost instinctive distaste merely
reinforced the deeper issue raised by the claim that the earliest humans
had evolved from an animal ancestry. Bishop Samuel Wilberforce's attack

on Darwinism at the 1860 meeting of the British Association raised this issue, prompting Huxley's response that he would rather be descended from
an ape than from someone who misused his position to attack a theory he
didn't understand. This continued to be an issue into the early twentieth
century. Not for nothing did the trial of John Thomas Scopes for teaching
evolutionism in Tennessee in 1925 become known as the "monkey trial."
Both of these episodes have achieved iconic status as symbols of the inevitable conflict between science and religion, although historians now recognize that their symbolism rests on manipulations of our perception by
extremists on both sides, each seeking to mythologize the past to their own
advantage.[7]

The problem was that Christianity has traditionally assumed that only
humans have souls. Unlike animals, our personalities have a spiritual component that will survive the death of the body and be judged by our Maker.
How could such an entity appear in the world through a natural process
starting from a purely animal ancestry? Conservative Christians insisted
that it could not, and rejected the whole idea of evolution. If the human
species had to be a new creation, then all the animal species might as well
be seen as products of supernatural intervention too. The only possible
compromise would be to limit the evolutionary explanation to the body
alone, calling in the supernatural to explain the abrupt introduction of the
soul into the first true human. Alfred Russel Wallace accepted this position, in part because he had become convinced by spiritualist phenomena
that the soul did indeed survive the body.[8] It remains the formal position of
the Roman Catholic Church. In the early twentieth century, the psycholo-
gist Conwy Lloyd Morgan and others proposed the concept of "emergent
evolution" in which the appearance of a totally new level of reality such as
the human spirit was somehow triggered when natural evolution reached a
certain level of complexity. This does at least avoid the need for a miracle,
but supposing that a Creator builds such discontinuities into the evolutionary process doesn't really satisfy most scientists.

Eventually many evolutionists who wished to retain some form of religious belief ended up endowing animals themselves with a primitive mental capacity, capable of being upgraded to the level of the human spirit. Evolutionists could then turn their backs on materialism by seeing the ascent of
life as the triumph of mind over matter. They eliminated the gulf between

animal and human not by turning humans into brutes but by uplifting the whole of animal creation to a quasi-human level.

The Connection Broken

The ape connection reinforced the theological problem, at least for the general public, who saw apes and monkeys as disgusting, gibbering creatures that seemed most unlikely to have improved themselves into spiritual beings. If these were our closest relatives, then the evolutionary connection was rendered all the more implausible. The developmental model of evolution did, however, offer a way out of the dilemma. In its simplest form, this model implied a linear ascent from ape to human via an intermediate "ape-man" stage colloquially known as the missing link. But if there were indeed predetermined developmental trends built into primate evolution, the theory of parallelism offered a way of severing the direct link between ape and human. Perhaps the similarities that were commonly seen as evidence of divergence from a recent common ancestor had actually been acquired independently by two separate evolutionary lineages, each subject to a similar predisposition. In the theory of parallelism, humans and apes might be derived from stocks that had been separate from the dawn of primate, or even mammalian, history. The human line had never passed through an ape stage, and its true ancestors had always had a higher mental and moral potential.

Such a theory of human origins was actually proposed in the early twentieth century by Frederic Wood Jones in Britain and Henry Fairfield Osborn in America.[9] Wood Jones derived humanity from the spectral tarsier, a cute little creature that already seems full of potential intelligence. Osborn postulated a bipedal ancestor inhabiting central Asia, long separate from the tree-dwelling apes of southern Africa. This idea was quite explicitly used as a means of countering the antievolution rhetoric that had become prevalent among fundamentalists by the 1920s. It inspired the expeditions that led to the discovery of the remains of "Peking man" in China (now included in the species *Homo erectus*). In its most extreme forms the theory of parallelism did not catch on, but there was a general assumption in the early twentieth century that the human lineage was geologically ancient and had not originated in Africa. This explains why the first fossil of

Australopithecus, discovered in South Africa in 1924, was initially rejected as a clue to human origins. We now regard the Australopithecines as the first hominids to emerge from an ape ancestry, but this interpretation only became popular in the middle decades of the twentieth century, coinciding with the emergence of the neo-Darwinian synthesis.

Curiously, the palaeontologist who put the Australopithecines on the map as human ancestors, Robert Broom, was a deeply religious man who thought that the whole sweep of evolution up to humanity revealed a divine plan. In general, the ape link was a barrier to acceptance of evolutionism by religious thinkers, and the theories of parallel evolution were intended to remove that barrier. But imagine what might have happened in a world where there was no Darwin to throw his weight behind the idea that humans and apes shared a common ancestor. If evolutionism in the late nineteenth century had emerged in a developmental form with a greater component of parallelism, the Wood-Jones/Osborn thesis might have been proposed earlier and would have carried even more weight than it did in our own world. In these circumstances it would have been less easy for the religious opponents of evolutionism to play upon the public's fear of an ape ancestry in order to reinforce their claim that the human soul could not have emerged by this route.

ORDER AND HARMONY

All the evidence from our own world tells us that liberal Christians found it easy to overcome their misgivings about the status of the human soul, provided they could believe that the evolutionary process was designed to produce the human species as its ultimate goal. Indirectly, at least, this would preserve our unique status in the world. But most people only felt comfortable viewing evolution as a purposeful and progressive force that operated according to laws built into it by the Creator. Darwin's theory of natural selection threatened this potential compromise on several fronts. Allowing the struggle for existence to weed out the few favorable variants accidentally produced by random variation didn't look like the kind of process a wise and benevolent Creator would employ. Darwinism threatened religion because it seemed to eliminate any form of teleology or purposefulness in nature. Darwin tried to head off this perception in the conclusion to the *Origin of Species* by suggesting that in the long run, at least, natural selection would gradually lead to progress. But his vision of progress was so haphazard and unstructured that it didn't convince many of his readers that it would suffice to maintain the necessary element of design.

The gradual conversion of the liberal Christian churches to evolutionism almost invariably entailed the sidelining or actual rejection of the theory of natural selection. Religious thinkers could accept evolution only if it was seen as purposeful and progressive, and this meant bringing in supplements or alternatives to selection. This is why the developmental worldview and the associated non-Darwinian mechanisms of evolution were increasingly substituted for Darwinism, to the extent that by the end of the century it was widely believed that Darwinism was on its deathbed.

In a world without Darwin, the absence of the selection theory in the early decades of the debate would have made it much easier for religious thinkers to accept evolutionism. Developmental theories were much simpler to reconcile with the traditional vision of cosmic teleology. The alternatives to Darwinism actually tried out in our own world would have been the vehicles through which evolutionism was presented to the public for the first time. The movement begun by Chambers's *Vestiges* in the 1840s would have continued uninterrupted by the crisis presented by the *Origin of Species*. Scientists would have gradually begun to support evolutionism over the course of the 1860s, but the developmental approach would have played a more significant role in their conversion. The supporters of materialism and scientific naturalism would have lacked one of their most potent arguments against religion. Spencer and Haeckel would still have been evolutionists, of course, but their systems would have been even more clearly based on the Lamarckian mechanism. Some, including T. H. Huxley and John Tyndall, may well have focused more of their attention on other areas of biology such as the latest developments in physiology and the neurosciences. Tyndall was, in any case, a physicist whose real inspiration was the idea of the conservation of energy (which left no room for supernatural action). With no debate on natural selection to distract them, religious believers would be in a better position to see how nonselectionist mechanisms of evolution could be used to preserve the element of design they needed.

Paley Transcended

Darwin presented his theory as a challenge to William Paley's concept of design. His natural selection replaced divine benevolence as the explanation of why species are adapted to their environment. But this was never the whole story, because any theory based solely on adaptation still tends to leave nature looking like an ad hoc collection of individual relationships between organisms and their environments. Much of the resentment against natural selection was based on the feeling that a wise Creator would construct His universe according to a rational plan rather than a jumble of local arrangements. A process of adaptation might generate structural improvements and thus lead to a form of progress, but it could not ensure progress in a single direction toward a predetermined goal. The tree of life depicted in the one diagram Darwin included in the *Origin* had no central trunk and no predetermined trends, and this would be true for any theory based on local adaptation, not just for natural selection. Sir John Herschel's complaint that natural selection was just the "law of higgledy-piggledy" could be applied to any theory that left outcomes to the mercy of local conditions— the element of random variation just made the implication more obvious. If the hazards of migration and local environmental fluctuations determined the outcome for each species, there could be no overall plan or pattern of creation. Even if natural law governed every element in the scheme, the fact that there was no prearranged coordination between those elements would mean that the outcome was irregular and haphazard.

Herschel's complaint reflected dissatisfaction with any worldview that does not allow the world to be seen as an expression of an underlying ordering force or pattern. It would apply almost as much to Paley's vision of the world as to Darwin's, since Paley's also made it impossible to see God as a source of order. For many nineteenth-century naturalists, as for many religious believers, it was important to be able to see an ordering principle in the world, a principle that revealed the mind of a rational Creator at work beneath the apparent hurly-burly of the everyday world. This attitude was fundamental to the structuralist or formalist tradition in nineteenth-century biology, especially prevalent in Germany. It was imported into Britain during the 1840s by Richard Owen, who insisted that the underlying unity of nature indicated by the vertebrate archetype was better evidence of creative design in nature than all the individual cases of adaptation cited by Paley and his followers. The human form might be the highest manifestation of the Creator's powers, but it was based on the same fundamental plan as any other vertebrate species. The developmental program in late nineteenth-century evolutionism can seen as an extension of this formalist position, an attempt to see the history of life as the unfolding of a coherent plan or pattern that could be interpreted as evidence of design. Without Darwin's intervention, this approach would have dominated evolutionism and made the movement far more attractive to religious thinkers.

In an article surveying the British contributions to this tradition, I called it the "idealist" argument from design. Critics have taken issue with that term on the grounds that Owen's appeal to a kind of Platonic idealism was only skin deep and that some exponents of developmentalism were neither philosophical idealists nor particularly religious.[10] But my analysis was meant to depict the biological debate in terms of what John Stuart Mill identified as the main fault line in British thought at the time, between Jeremy Bentham's utilitarianism and the idealism of Samuel Taylor Coleridge. Bentham's political philosophy praised actions that were useful in promoting human happiness, and Paley's focus on adaptation as evidence of divine benevolence was an application of the same principle. Coleridge and the Romantics despised the practical emphasis of the emerging Industrial Revolution and sought the meaning of life in deeper spiritual realities. By using the term "idealist" to denote the structuralist alternative in biology, I was suggesting that it lies firmly on the Coleridgian side of this dichotomy, because the unity and harmony of nature can be seen as expressions of the divine mind at work. I agree that not every exponent of structuralism made this connection, and I have argued that the functionalist mindset represents something more akin to a personality type.[11] There are some people who just can't bear to think that they are living in a world that is fundamentally disorderly. They instinctively react against the suggestion that something as trivial as local adaptation can be the crucial factor shaping the natural world. This mode of thought often expresses itself through the conviction that some kind of rational ordering principle, most obviously derived from the mind of God, imposes an underlying unity on the apparent diversity of life.

Historians of German science note that people of that country readily accepted the structuralist tradition as being compatible with religious belief. German transcendentalists such as Lorenz Oken were explicitly idealist in their approach (significantly, Owen arranged for a translation of Oken's work into English). Transcendentalism offered an alternative to the native utilitarian tradition in Britain, and although many religiously minded naturalists were suspicious at first, they gradually warmed to Owen's efforts to convince them that there was more to be seen in the argument from design than Paley had suggested. As evolutionism became more popular in the 1860s, there were numerous efforts to suggest that not every characteristic could be seen as having adaptive significance. W. B. Carpenter, by no means an opponent of Darwinism, noted the beauty and harmony displayed by the Foraminifera shells he studied. The Duke of Argyll and the anatomist St. George Jackson Mivart emerged as supporters of theistic evolutionism, both presenting evidence that evolution could not be seen as a purely adaptive process. Argyll argued that the colors displayed by many species, especially birds, indicate that the Creator intended to produce a beautiful world, while Mivart appealed to evidence of parallelism to undermine the case for purely divergent, adaptive evolution. Mivart was a disciple of Owen, and we have seen how their school of thought became the springboard for the emergence of developmental evolutionism. Theirs was a school of thought explicitly intended to present evolution as a theory compatible with religion.

Orthogenesis and Design

The theories of non-adaptive orthogenesis that emerged in the late nineteenth century were an extension of the formalist position, although when pushed to this extreme, there was less room for synergy with religious faith; rigid trends driving species toward racial senility and extinction could seem almost as threatening as random variation and selection. Yet there was surprisingly little emphasis on this potentially darker side of developmentalism. For example, Henry Fairfield Osborn, the leading American exponent of orthogenesis, wrote an extremely upbeat account of the spiritual progress made possible through evolution.[12] Racial senility always occurred along the side branches that diverged from the main line of advance leading toward humankind. Robert Broom sought to demonstrate that humanity was the intended goal toward which evolution had progressed by showing that all other branches of evolution had diverged into overspecialized dead ends.[13] This preserved the essential optimism of the broad developmental program encapsulated in the image of a main line of evolution progressing steadily toward humankind. In the non-Darwinian world, this linear progressionism would have been even more prominent, rendering the evolutionary movement much less of a threat to religion.

PROGRESS AND PURPOSE

But what about the relationship between an organism and its environment? This could not be ignored in any comprehensive evolutionary theory, least of all in a culture where Paley's version of design was still taken very seriously. For all his efforts to unify nature via the archetype, Owen knew that the divergence of groups away from a common ancestor or archetype represented the unfolding of various adaptive specializations. This is why Darwin could use Owen's interpretation of the fossil record as evidence. But Darwin's explanation of the trends as a consequence of natural selection was widely thought to undermine any hope of seeing adaptation as an indication of divine benevolence. And how could a theory based on divergent trends become the foundation for an overarching vision of progress toward a morally significant goal?

In the conclusion of the *Origin* Darwin suggested that, in the long run, natural selection would lead to the evolution of higher types of organization. Most of his followers believed that evolution was inherently progressive, but accepted that it was not possible to see the development as a single line of ascent leading directly toward a predetermined goal. The logic of developmentalism had to be tempered by recognition that progress is superimposed onto a system of localized interactions between organisms and the ever-changing environment. Darwin's focus on the complexity of what would later be called ecological relationships, expressed in his metaphor of the "tangled bank," offered a vision that had its own power to inspire. We see this in the use of evolutionary metaphors by literary figures, and the same inspiration is still used today to counter the creationists' anti-Darwinian rhetoric.[14]

That rhetoric has a long history, however, stretching back to Wilberforce's attack in 1860, and the specter of a completely amoral nature driven by brutal struggle certainly haunted the late nineteenth-century imagination. Without Darwin's intervention, there would have been alternative, less threatening ways of imagining how progress and purpose could be built into the natural interactions between organisms and the environment. We know this because those alternatives actually flourished in our own world, where they played a major role in limiting the damage that Darwinism did to the relationship between science and religion. In a world without Darwin, these ideologies of progress would have ensured that evolutionism had a moral foundation that most religious believers could feel comfortable with.

Theistic Evolutionism

When the Harvard botanist Asa Gray, a staunch Presbyterian, sought to present Darwinism as compatible with his religion, it was the threat to Paley's argument from design that offered the greatest obstacle. Gray tried to argue that since the Creator instituted the laws of nature, it didn't really matter how those laws operated as long as their end result was to maintain a species in a state of adaptation. God's benevolence was ensured whether He produced useful characteristics directly by miracles or indirectly by law. But in the end Gray found it hard to sustain this argument if the mechanism in play was natural selection. How could the constant elimination of a host of useless variations—he called them "the scum of creation"—be seen as an expression of divine benevolence? He eventually conceded that it would be better to assume that "variation has been led along certain beneficial lines."[15] Darwin protested that this would make selection superfluous: if God has designed the laws of variation, one would have to believe that He somehow preordained every trivial characteristic, including the shape of one's own nose.

Gray's dilemma drove him toward what has been called theistic evolutionism, the claim that the course of development cannot be explained unless divine foreknowledge is built into its underlying laws. But how could God build foreknowledge of every twist and turn in a species' environment over the course of geological ages into the laws governing variation and heredity? Unless one were to suppose that God perpetually interfered with the normal process of variation, there would have to be an indirect way in which an organism's needs were made known to the internal processes governing its reproduction. There was, in fact, such a process already under discussion, although Gray does not seem to have realized its potential: the Lamarckian theory of the inheritance of acquired characteristics.

Rival Visions of Lamarckism

Lamarckism has acquired a reputation as a theory intrinsically hostile to materialism. Thinkers appalled at the moral and spiritual implications they associated with natural selection certainly used Lamarck's theory as an alternative to Darwinism. Writers such as Samuel Butler and George Bernard Shaw hailed it as the key to an evolutionism that would once again allow the world to be seen as the expression of moral purpose. Lamarckism does not depend on the struggle for existence (although struggle could be seen as a spur that would encourage individuals to improve themselves). Instead, it is a mechanism based on nonrandom, purposeful variations, a process by which animals develop new characteristics through new habits—which is, of course, an adaptive response to environmental change. Its only problem is the lack of hard evidence that such individual modifications are actually transmitted to the next generation. But in the late nineteenth century, most naturalists, Darwin included, thought that there was enough indirect evidence to make the theory plausible. In the 1870s, it emerged first as a major supplement to natural selection and eventually as a complete alternative to the whole Darwinian program. In a world without Darwin, Lamarckism would have been promoted as the most plausible explanation of adaptive evolution and would have been a central plank in any evolutionary theorizing.

Lamarckism played a key role in the thinking of two of the greatest popularizers of evolutionism, Herbert Spencer and Ernst Haeckel. I have argued that both would have been able to promote their evolutionism without the spur provided by Darwin, and in the counterfactual world, Lamarckism would have been the centerpiece of their program. But here we encounter an apparent paradox. Spencer and Haeckel are seen as exponents of just the kind of naturalistic worldview that Butler and Shaw invoked neo-Lamarckism in order to reject. How can the same evolutionary mechanism have been a component of both the naturalistic philosophy and its alternatives? There are two ways of answering this question. The first is to point out that there is no one-to-one relationship between scientific theories and philosophical or moral positions. Different aspects of a theory can be spun in different ways, although one manifestation sometimes becomes so popular that another is forgotten.

The second approach is to recognize that ideological polarizations are seldom as rigid as the extremists would like us to believe. Even in our world, there were religious thinkers who welcomed Spencer's evolutionism, so we must be very careful not to assume that his writings were only interpreted in a naturalistic light. Darwin's theory has almost certainly played a major role in shaping later perceptions of both Spencer and Haeckel. By accepting natural selection (even on their own terms), they allowed later generations to perceive them as exponents of a completely antireligious worldview. Darwinism became the symbol of materialism, so any thinker associated with it in the public mind became by definition a materialist.

But neither Spencer nor Haeckel were materialists, and in the 1860s and 1870s their writings were often read in a way that was sympathetic to liberal religious beliefs. This was what actually happened in our world, and it was only toward the end of the century that the rival perception came to the fore, allowing Lamarckism to be claimed exclusively by the moral reaction against materialism. In a world where Spencer and Haeckel were not tempted by the Darwinian alternative, the Lamarckian component of their thought would be more clearly apparent and its appeal to religious thinkers would be unmistakable—even to the next generation. We have been conditioned to see these thinkers solely as opponents of religion, but that is a simplistic dichotomy forced on us by later critics. In the non-Darwinian world, the selection theory would not be there to serve as a symbol of what the later generation reacted against. These later critics would find it harder to dismiss their predecessors as simple materialists—and harder to pretend that their support for Lamarckism represented a new scientific inspiration.

Naturalistic Lamarckism

Herbert Spencer had been promoting evolution as the key to a new worldview since the early 1850s. His was certainly a naturalistic philosophy in which the everyday interactions between individuals and their environment accumulated over vast periods of time toward higher mental and moral faculties. Lamarck's mechanism was the key that opened the door to this new vision, and Spencer only belatedly absorbed Darwin's theory of natural selection into his system. Natural selection was an indirect process for bringing a species into equilibrium with its environment, but the direct involvement of organisms themselves in developing new habits and characteristics was far more important. Spencer simply couldn't believe that such purposefully acquired characters were not transmitted to future generations so as to improve the species as a whole. He probably saw natural selection as a mechanism that eliminated those individuals who did not have the capacity to adapt themselves to changes in the environment.

Spencer's philosophy was a product of the new liberal ideology of free-enterprise individualism that was driving the industrialization of the economy. It was naturalistic in the sense that it offered no way of seeing beyond the phenomena of the material world. But Spencer, like T. H. Huxley, was anxious to avoid being branded an atheist. Huxley coined the term "agnostic" to define the view that we have no means of determining if there is anything beyond the natural world, but neither he nor any of his fellow agnostics wanted to suggest that we could positively know there was no God. Indeed, many believed that there almost certainly was a higher reality beyond the material world—they just denied orthodox religion's claims to have access to that reality. Spencer called this higher reality the "Unknowable," giving religious believers a bridge over which they could cross to reconcile the new liberalism with traditional beliefs. His evolutionism also saw a goal toward which progress was aimed: a perfect society in which all conflict would disappear (a profoundly non-Darwinian perspective). We know that there were many Protestant clergymen who willingly crossed that bridge and saw Spencerianism as the key to a second reformation of Christianity, bringing it once again into line with an ongoing moral transformation of society. Far from fearing Spencer's philosophy as a threat to moral values, they saw it as offering a reformulation of morality in response to new social developments.

By allying himself with Darwin, Spencer opened himself up to claims by more conservative Christians that he left everything to be settled by a brutal struggle in which only the most rapacious would survive. But that was never his intent, and in a world without Darwin, the conservatives would have been deprived of their most potent means of misrepresenting his system. The image of Spencer as a social Darwinist still haunts our imagination today, but the next chapter will show that this is a myth rather than a true picture of his system. Spencer's evolutionism had a clear ethical dimension, and Lamarckism was the key to his understanding of how higher mental and moral faculties were developed.

It was this moral dimension that allowed liberal Protestant clergymen to accept progressive evolution as the means the Creator used to form the moral and spiritual character of humankind. James Moore calls these clergy "Herbert Spencer's henchmen," and in a world that was not exposed to the challenge of Darwinism, this group might have formed a body of opinion even more influential than the naturalistic Spencerians.[16] The alliance was made possible by the fact that Spencer's evolutionary explanation of the moral sense focused on the enhancement of what has been called the Protestant work ethic. This approach to morality sidelined (although it did not ignore) charity in favor of the more self-sustaining virtues of thrift, industry, and initiative. Spencer allowed a role for competition but saw it as the spur to individual self-improvement, with improvement understood as a balance between material success and moral character. The system was evolutionary because Lamarckism allowed the individual acts of self-improvement to accumulate over generations and thus mold the character of a species. It was easy for a liberal Christian to see this as a system that could have been put in place by the Creator to achieve the production of the human race. Lower down the scale, of course, it also adapted species to their environments, but in the long run these local adaptations would generate new levels of physical and mental capacity.

The Anglican clergyman Charles Kingsley had already broached this mode of reconciliation, probably without direct contact with Spencer's writings. Kingsley's story *The Water Babies*, usually portrayed as a response to Darwin, was in fact an expression of a thoroughly Lamarckian vision of progressive evolution in which individuals, races, and species all progress to higher levels of moral capacity if they respond positively to challenges from the environment, but degenerate if they take the easy way out. Darwinians could certainly recognize this implication—Alfred Russel Wallace and E. Ray Lankester being clear examples—but for Kingsley there was a direct transition from individual self-improvement to the progress of the race, and this was a Lamarckian, not a Darwinian, view of the process. In effect, Kingsley was proposing a muscular Christian version of Spencerianism, and as Spencer's philosophy became more popular, liberal Christians everywhere began to see the connection. In America, John Fiske's *Outlines of Cosmic Philosophy* of 1874 presented Spencer in a form sympathetic to religion. Clergymen such as Henry Ward Beecher emerged as enthusiasts for Spencerian evolutionism. As Beecher admitted, the move required a significant reformulation of Christian principles, most obviously the abandonment of the traditional notion of original sin. Spencerianism suggested that humans had risen to their present state, not fallen from a higher one, and Christ's ministry had to be seen as a call to throw off the legacies of our animal ancestry. But for many this seemed a price worth paying to maintain Christianity's role as a moral force in a rapidly changing world.

In a world that was not troubled by the harsher vision of evolution promoted in the *Origin of Species*, this hybrid of liberal Christianity and Spencerianism would have been the dominant form of evolutionary thinking during the 1860s and 1870s, and it would have remained influential in America long after that. For Spencer himself, natural selection was something of a distraction, and he would have been quite happy to push his philosophy forward without it. And without the Darwinian theory of selection, his vision of struggle as the spur to individual and racial self-improvement would have been an even clearer beacon to religious thinkers. Scientific naturalists such as Huxley might have stood aside (Huxley was never an enthusiast of Lamarckism) and focused more of their attention on other areas of science that offered clearer support for their cause, most obviously the latest developments in physiology and the study of the nervous system. Evolution would certainly be an arena for debate between the two sides, but without Darwin it could not be so easily tarred with the brush of extreme materialism. Conservative Christians would still reject it, of course, but their hostility would be muted both by lack of a key argument and by the even more visible support for the vision of progressive development among their fellow believers. Many historians now believe that we have exaggerated the influence of Huxley and the scientific naturalists of the period, partly as a result of the success of their own propaganda.[17] In a world without Darwin, their failure to dominate the cultural debate would be even more apparent.

In Germany, Haeckel's evolutionary philosophy was even more virulently anticlerical than Huxley's, yet here too there were elements that could be exploited by those seeking a less divisive reception for the new science. Haeckel was not a materialist; his monist philosophy taught that mind and matter are different aspects of a single reality, so that all living things—indeed all material objects—had a mental component. The human mind was the end product of a long sequence of mental progress. Haeckel's recapitulation theory also provided a model for those who wanted to see that progress as a predetermined advance toward maturity. Like Spencer, Haeckel was a Lamarckian who saw most interactions between an organism and its environment as progressive and purposeful. Finally, he openly presented aspects of his teaching as continuations of the transcendental worldview popular earlier in the century—his evolutionary heroes were Darwin, Lamarck, and Goethe, and in the world without Darwin, there would only have been the latter two. Here, from a background very different from Spencer's, was a progressionist vision that could be seen as the unfolding of some ultimate moral purpose. Haeckel's monist philosophy, with its implication that mind might be seen as the driving force of evolution, anticipated the turn that Lamarckism would take toward the end of the century.[18]

Neo-Lamarckism and Creative Evolution

Critics of Darwinism who became vocal in the decades around 1900 have shaped our perception of Lamarckism. Beginning with the novelist Samuel Butler, people of note have vilified Darwin's theory as the expression of a ruthless materialism, while presenting Lamarckism as its worthy alternative. This polarized image of the relationship between the two ideologies was taken up in the early twentieth century by writers such as George Bernard Shaw, who allied it with the vitalist philosophy of Henri Bergson. Bergson's vision of a creative life force struggling to overcome the limitations of the material world defined a new evolutionism that was anxious to present itself as transcending the old Darwinian materialism.[19]

In the process, the fact that Spencer and Haeckel had also been Lamarckians was conveniently forgotten, along with the fact that Spencer's philosophy had been well received by many religious thinkers. The myth of a late nineteenth century completely dominated by Darwinism and materialism established itself in the public mind, allowing the new Lamarckism to be seen as the means by which traditional moral values would be revitalized. But this rewriting of history would not have been possible in a world that had not been exposed to the trauma of the original Darwinian debate. Spencer and Haeckel would not have been tempted to include the element of selection in their otherwise progressionist philosophies. Without the specter of ruthless Darwinism looming over the first generation of evolutionists, the moral component of developmentalism and Spencerianism could not have been so easily denied, and the continuity between the old Lamarckism and the new would have been unmistakable. Even if the selection mechanism had been recognized in the early twentieth century, it would have been incorporated into an evolutionism that was far more securely identified with liberal religion than with the more aggressive forms of naturalism.

Samuel Butler's *Evolution Old and New* of 1879 was perhaps the first shot in the campaign to rehabilitate Lamarckism, a mechanism that had itself been seen as an agent of materialism earlier in the century. In this and a series of later books, Butler attacked the selection theory for its implication that animals (and thus humans) are merely puppets at the mercy of an implacable material world, condemned by their heredity either to death or to temporary survival won by brute force. He hailed Lamarckism as a theory that offered a way out of this nightmare for moralists and religious thinkers who valued the traditional view of humans as moral agents. Butler was recreating Lamarckism as a force against naturalism, but he depicted his opponents not in Spencerian but in the most materialistic of Darwinian terms. There was no mention of the key role that Lamarckism played in Spencer's evolutionism, presumably because Butler found it easier to vilify an oversimplified Darwinism than to explore the more complex issue of how his version of the theory differed from that already in play. There was a real difference: for Spencer and Haeckel the adjustment of the organism to changes in its environment was a more or less automatic process, whereas for Butler it involved a deliberate choice of new habits. But Butler preferred to see himself as the pioneer of a new, more moral evolutionism, not as someone who was merely putting a different emphasis on an existing idea.

Butler launched personal attacks against Darwin, claiming he had ignored his predecessors' contributions, and the Darwinian community subsequently ostracized Butler. But even in our own world, nonselectionist theories were actively explored in the 1860s and 1870s, and they became more popular as the century progressed toward its close. The scientific community, including Darwin's son, Francis, eventually began to take Butler more seriously. Religious thinkers such as Henry Drummond and Reginald Campbell hailed the emergence of a "new theology" based on non-materialistic evolutionism. In America, writers such as Edward Drinker Cope and Joseph LeConte proposed similarly non-materialistic approaches.

Cope proposed an essentially vitalist form of Lamarckism in which organisms drew on a nonmaterial growth force to shape their future evolution. The weakness in Cope's position is that only the founder members of an evolutionary branch were free to choose a new lifestyle—all their descendants were locked into a well-defined trend of specialization. But generations of enthusiastic opponents of materialism ignored the link between Lamarckism and orthogenesis, being anxious to endow living things with mental powers that could be seen as an expression of divine creativity.

The playwright George Bernard Shaw was no enthusiast for formal religion, but he was a passionate opponent of materialism, and he identified Darwin's theory as a major source of the attitudes and values he abhorred. He famously declared that if natural selection were true, "only fools and rascals could bear to live."[20] He linked his endorsement of Butler's Lamarckism (not entirely accurately) with Henri Bergson's philosophy of "creative evolution." Bergson's vision of a creative life force, the élan vital, struggling upward against the inertia of brute matter was not explicitly Lamarckian, but it did encapsulate a new vision of progress in which life's evolutionary adventure was seen less as the unfolding of a predetermined pattern and more as an exploration of the possible. His book *Creative Evolution* was translated in 1911, and its more open-ended view of progress inspired many biologists, including some of those who reformulated the idea of natural selection in the 1920s and 1930s. Religious thinkers, anxious to jump on the bandwagon of anti-materialism without abandoning some of the liberalizing gains of the previous century, also took it up with enthusiasm.

Shaw's focus on Darwinism as a symbol of everything that was wrong with late nineteenth-century materialism encouraged a rewriting of history that has distorted our perception to this day. This revisionism reemerged, for instance, in Arthur Koestler's attempt to rehabilitate the Lamarckian experiments of Paul Kammerer.[21] People saw the vitalist form of Lamarckism as something entirely new, a reaction against the dead hand of materialism ushered in by Darwin. There was no recognition of the substantial resistance to Huxley's naturalistic philosophy, even in the scientific community, and no recognition that Lamarckism had been a vital part of evolutionism from the start. What emerged was a caricature both of late nineteenth-century culture in general and of evolutionism in particular. Only in the last few decades have historians begun to rescue the true story of non-Darwinian evolutionism and tell the more complex tale of its role in the relationship between naturalism and belief.

Now consider what might have happened in a world where there was no selection theory for Spencer and Haeckel to incorporate into their evolutionary thinking. The developmental version of evolutionism would have held even greater sway both in science and in general thought. Naturalism would have been robbed of one of its most potent arguments, and it would be easier to see Spencer's and Haeckel's progressionism as a revision of the traditional religious worldview, rather than as a challenge to it. There may well have been a reaction that placed new emphasis on life as an active force in nature, but it would have been harder for writers such as Butler and Shaw to present their Lamarckism as a totally new initiative. They would have had to admit that they were merely reconfiguring a developmental system that was already established as the foundation of evolutionary thought. Even when the selection theory was eventually developed, it would have been seen as yet another theme to build into the developmental story, not as a threat to the whole system. There would be much more continuity in the process by which moral philosophy and liberal Christian thought absorbed the evolutionary message and no Darwinian bogeyman haunting the imagination of successive generations.

THE FUNDAMENTALIST RESPONSE

One voice that has not yet been articulated in the debates surveyed so far is that of conservative Christians, especially those from the evangelical tradition. Here there was deep suspicion of the whole ideology of progress because it undermined the traditional Christian view of the human situation based on sin and redemption. Evangelicals also held a much deeper reverence for scripture. In their view, the liberals' efforts to strike a compromise with modern ideas was a rejection, not a reformation, of their faith's core foundation.[22] This attitude must have persisted throughout the period we see as dominated by the ideology of progress, but it was seldom articulated in the circles where evolutionism was debated. In Europe, the evangelical influence steadily declined, but in America the rise of what became known as fundamentalism heralded a resurgence of evangelical fervor based on a rejection of modernizing cultural trends that were feared as a threat to

traditional family values. Darwinism became a symbol of declining morality, and the result was a move to resist the intrusion of evolutionism into
the educational system. The trial in 1925 of John Thomas Scopes for teaching evolution in Dayton, Tennessee, in turn has become a symbol of the
fundamentalists' determination to reject the scientific interpretation of human origins.[23]

But did it have to work out this way? The rise of fundamentalism was
almost certainly inevitable. It represents a broad social movement that
could hardly have been deflected by the absence of a single scientific idea.
But was it inevitable that evolutionism became the symbol of everything
the fundamentalists distrusted about modern thought? Darwinism became
that symbol because it could be used to present evolutionism in its most
materialistic form. If there had been no debate over natural selection in the
late nineteenth century, the developmental version of evolutionism taken
on board by liberal Christians would have been widely accepted as an ideology of purposeful development toward a morally significant goal. This does
not mean that the fundamentalists would have welcomed progressionist
evolutionism, since it was the idea of progress itself that they distrusted.
But without the specter of Darwinism to haunt their imagination, it is possible that they might not have chosen to pick a fight on this issue. The campaign against modernism would have had other targets, if only because this
would have minimized the potential for conflict among the liberal and conservative wings of the Christian community.

The pamphlets known as *The Fundamentals* were published between
1910 and 1915 and are widely supposed to represent the start of the renewed campaign against evolutionism. In the 1920s, Americans in several
Southern states resisted efforts to introduce evolution into the school cur-
riculum, culminating in Tennessee's Butler Act, under which Scopes was
prosecuted. But modern historians have shown that much of the popular
understanding of this episode is based on myths that grew up around it in
later decades and that were summed up in the play and subsequent movie
Inherit the Wind.[24] In fact, some of the authors who wrote for *The Fundamentals* were not opposed to non-materialistic evolution.[25] And while William Jennings Bryan, the politician who helped to prosecute Scopes, focused the Fundamentalists' attention on evolutionism, he did not himself
believe in a literal interpretation of Genesis. By no means did all Southern
states pass laws against the teaching of evolution, in part because the evo-

lutionism that was taught in schools was carefully shaped by the scientific community to avoid presenting it in a Darwinian light. The location of the Scopes "monkey trial" was determined partly by Tennessee politics and the renewal of a school textbook contract, and partly by the desire of the citizens of Dayton to put their town (which had run into economic difficulties) in the national spotlight.[26] Far from being an inevitable expression of a cultural conflict, the whole episode is shot through with the sort of contingencies that allow counterfactual possibilities to be imagined.

This is not to imply that the evangelicals would have been comfortable with evolutionism. The link between apes and humans would still have been offensive, and concern for the traditional Christian message would have made evangelicals suspicious of efforts to undermine the idea that humans were divinely created. But in a world where evolutionism was routinely portrayed as purposeful and progressive rather than undirected and based on brutal struggle, it would have been less of a priority as evangelicals sought to identify the sources of the modernizing trends that were undermining traditional values. Even within science, there were other theories that could be seen as a threat to human dignity and spirituality, ranging from materialism in the biomedical sciences to the new initiatives in psychology. Sigmund Freud's analytical psychology was actually a belated product of the recapitulation theory, and was thus a distorted expression of the developmental paradigm.[27] But because Freud concealed the evolutionary element in his theory to stress his own originality, it would have been seen as a truly revolutionary vision of the darker aspects of the mind's animal foundations. It would probably have emerged in the non-Darwinian world and might have attracted a bigger share of the evangelicals' ire. Some even saw Einstein's theory of relativity as a threat to moral values. In a world without Darwin, there would have been plenty of other targets for the fundamentalists to identify as the chief threats to their Christian heritage.

All this takes us a long way past the point of divergence between our own world and the counterfactual one I am trying to imagine. So I do not offer these brief speculations as a serious attempt to visualize a situation in which evolution could avoid all conflict with religion. They are instead a challenge to make us think more carefully about the extent to which our perception of the alternatives now being debated has been shaped by history—and the manipulation of that history by later generations. If we can imagine a world without the Scopes trial, could we imagine one in which atheistic evolutionists and creationists were not at loggerheads in the early twenty-first century? With evo-devo as part of the story from the beginning, there would be less of a temptation for scientific materialists to focus on natural selection as a vehicle for undermining any vestige of purpose in nature, thus weakening the need for the alternative of intelligent design. The naturalistic tradition running from Huxley to Richard Dawkins would be robbed of one of its most potent arguments, making it a less threatening target for fundamentalists.

Can we also imagine a world in which Young Earth creationism did not become a central feature of fundamentalist thinking in the late twentieth century? Perhaps not—but the Young Earth position rejects the whole gamut of modern historical sciences, including cosmology, geology, palaeontology, and prehistoric archaeology. In a world where the theory of natural selection had emerged much later and in a much less confrontational manner, evolutionism would be just one of the targets, rather than the principal symbol of what the fundamentalists oppose. Perhaps Lyell would be seen as the real villain, since his uniformitarian geology could be seen as the springboard for the whole anti-biblical vision of earth history. This position was actually articulated in one movie made to coincide with the Darwin bicentenary year,[28] so it is not impossible to imagine it as the central plank of the creationist platform in a world where Darwin did not offer such an obvious target.

8

SOCIAL EVOLUTIONISM

William Jennings Bryan's attack on Darwinism focused as much on its moral as its religious implications. He had heard that German military leaders had launched their attempt to conquer Europe in the Great War partly because they believed that the struggle for existence between nations would determine which was superior.[1] This charge reappears in an updated form in the literature of modern creationism. Critics routinely portray Darwinism as the impetus behind dangerous social policies collectively known as "social Darwinism." Creationists frequently insist that without Darwin, there would have been no Great War, no Hitler, and no Holocaust. Nor is this claim without some academic support. Richard Weikart, a supporter of intelligent design, has argued that Darwinism was a key component in the belief system that led the Nazis to attempt the extermination of the Jews. In a more recent book, he is less specific, focusing more on the general biological component of Hitler's thinking, but he still identifies natural selection as an inspiration for the Nazis' drive to improve the human race and eliminate inferior types.[2]

That a generalized evolutionism might have found its way into the thinking of German militarists and later into Nazi ideology seems highly plausible. The idea of evolutionary progress stimulated by competition did become widespread and would have done so even without Darwin's theory of natural selection. But uncovering an element of evolutionary thinking in a ruthless political ideology is not the same as proving that the science actually created the attitude of cruelty and indifference. Other factors, scientific and nonscientific, were also involved, and counterfactual history will help us determine whether they could have provided alternative sources of inspiration and rhetoric. The more specific implication that Darwin's theory of natural selection was responsible for the consequences we

234 · CHAPTER EIGHT

deplore is even more open to question. I shall argue that in a world without Darwin, there would have been just as much social Darwinism—indeed, some aspects of it might have been even worse—it just wouldn't have had that name.

Leaving aside the question of whether we should reject a scientific theory solely because we don't like the social applications derived from it, how plausible is the claim that a single theory is responsible for creating so much human suffering? Do critics really believe that the science was so inspirational to evildoers that without its influence they would have taken a less maleficent course? Among all the complex social and cultural factors that led to the various horrors of the twentieth century, can this one theory have been the crucial trigger? Isn't it a little farfetched to imagine the German commanders (in either war) launching their assault because they were driven by a model of human affairs based on an out-of-date theory that had never been taken seriously by the scientific community? When the Great War began, people believed that scientific Darwinism was on its deathbed. It was barely being revived at the time of the Nazi atrocities, and even then, mainly in Britain and America. Even if leaders wanted to base their politics on scientific principles, wouldn't they prefer a theory that was both up-to-date and successful as science? Perhaps modern critics have forgotten how little impact the theory of natural selection actually had in biology until the mid-twentieth century. Or maybe they think Darwin's work inspired nonscientists despite its widespread rejection by the experts. But ordinary people get their information about science mainly from popularizations, and these seldom provide a completely accurate exposition of complex theories. This is why it is much more plausible to implicate a broad evolutionary progressionism, which would have emerged even without Darwin, as the scientific basis for these social developments.

Michael Ruse has argued that evolutionism became popular in science as a byproduct of the more general enthusiasm for the idea of progress in the nineteenth century.[3] Biologists were not converted by the strength of the scientific evidence—they adapted their theorizing to the general progressionism of the time. This explains the preference for non-Darwinian theories, and in a world without Darwin, the link between evolutionism and the idea of progress would be even more apparent. But Ruse's point exposes the weakness of any claim that the science was actually responsible for generating harsh social attitudes. If anything, the science was driven by general cultural developments, not the other way around. In our world, Darwin became the symbol of evolutionism, and later commentators picked out concepts associated with his theory of natural selection in order to highlight negative forces becoming apparent at the end of the century. Instead of recognizing that many of these alleged consequences would have emerged anyway as the idea of progress began to unravel, they used Darwinism as a scapegoat for the all the ills that had begun to plague society.

The opponents of Darwinism around 1900 insisted—not very convincingly, we have seen—that it had dominated science in the late nineteenth century. The phrase "social Darwinism" first came into use in the 1890s, and it was introduced as term of abuse. There have been few explicit calls by card-carrying social Darwinists to improve society through the ruthless elimination of the unfit. The earliest ideology identified as a form of social Darwinism was not militarism but the unrestrained free-enterprise capitalism espoused by the supporters of Herbert Spencer. Weikart accepts this as a form of social Darwinism,[4] but surely in so doing he exposes the paradoxical nature of his more visible claim that Darwin's theory helped to create Nazism. The paradox arises from the sheer diversity of the alleged derivatives of Darwin's theory. Spencer and his followers focused on individual competition, not national or racial struggle, and they were outspoken opponents of militarism. So how can the same theory have inspired mutually hostile ideologies? Surely if there is a true social Darwinism, it would be the version created by Darwin's own contemporaries, using the same intellectual and cultural resources. Paradoxically, this is precisely the ideology of free-enterprise individualism that many Americans still favor today—which ought to give creationists food for thought when they try to insist that Darwin promotes harsh social values.

DEFINING SOCIAL DARWINISM

In addition to militarism and ruthless capitalism, critics have identified other attitudes and values that are supposed to have been promoted by Darwinism. The idea that races represent fundamentally distinct human types, with some remaining closer to the apes, is often thought to derive from the theory. Yet we now know that Darwin opposed the most extreme racism of his time; it was primarily creationists and anti-Darwinian evolutionists who insisted that the races were distinct species. Eugenics, the belief that selective breeding can improve the human race, is another alleged spin-off (this one emerging several decades after Darwin published the *Origin*). But eugenics drew no analogy between society and natural evolution. It supposed that selection must operate artificially, just as it does in the long-established practice of animal breeding that Darwin himself used to explain how selective processes work. All of these complexities suggest that something much more complicated was going on—this is not a simple case of a new idea in science inspiring a single harsh ideology.

Selection and Society

In an attempt to clarify the situation, the historian of social thought Mike Hawkins has proposed five key components of social Darwinism.[5] He acknowledges that social Darwinism is not a single ideology, but more of a mindset highlighting factors that can be built into several different political systems. Two of Hawkins's components are very general and can apply to any form of social evolutionism. The first is the assumption that human actions are governed by natural law. This would apply to any system that bases human action on natural processes, including all materialistic philosophies. It would cover not just social evolutionism but also the much wider movement to see the human mind as subject to natural law. Long before Darwin, the phrenologists were teaching that the brain was the organ of the mind, and thinkers such as Spencer were exploring the materialistic implications of this view before Darwin published.[6] There had also been much controversy over the claim that social activity can be seen as subject to law, an idea advanced, for instance, by the historian Henry Buckle. At this level, social Darwinism simply ranks alongside the many other cultural influences driving a naturalistic way of thinking about humanity in the nineteenth century.

Hawkins's second point is more specific: social Darwinism assumes that the laws of biological evolution apply (or ought to apply) to human social development. But here too there are other forces at work. Darwin's was not the only theory of evolution proposed at the time, and a person could just as easily use the theories of any of his rivals as models for social progress. Hawkins's point does, however, focus our attention on the growing sense that the laws of natural evolution are somehow inescapable: if we do interfere with them, we upset the natural state of affairs, and the results will be disastrous. Hence the assumption that the "struggle for existence" is beneficial and should not be eased. But many in the late nineteenth century came to believe that modern civilization *had* suppressed the natural processes that governed the emergence of humankind from its animal ancestry. They were concerned about the possibility of degeneration, and the eugenics movement's call to apply artificial selection to the human race was the most obvious response.

Darwinists do not, in any case, have to believe that the laws of biological evolution offer us a model. Toward the end of his life, T. H. Huxley lectured on evolution and ethics, rejecting the claim that the "gladiatorial show" of nature should be seen as the model for human relationships. Paradoxically, Huxley had never really believed that natural selection was the primary agent of evolution, but now he used the image of nature's ruthlessness to distance human morality from our animal origins. His real target was Spencer's philosophy, which he now believed was responsible for widespread indifference to those who were unsuccessful in the struggle for life. Huxley in effect joined the ranks of those who maintained that with the emergence of the human mind, something new had appeared in the evolutionary process, a force no longer governed by the laws that had shaped previous episodes.[7]

Hawkins's final three points relate to the specific laws of evolution that the social Darwinist seeks to apply. These involve (1) the struggle for existence produced by population pressure, (2) the determinist view of heredity in which character traits are rigidly inherited, and (3) the mechanism of natural selection that, it is assumed, generates "fitter" species. The problem is that only his last point captures the real essence of Darwin's insight. The first two represent ideas that were widely disseminated at the time and that found their way into a variety of ideologies—and scientific theories— that were not Darwinian as defined by what biologists see as the core process of natural selection. There is also some ambiguity over what natural selection is supposed to produce. If it is merely new species adapted to the local environment, this seems hardly relevant to the human situation. But if progress toward higher levels of development is the desired outcome, we are already moving into the area where the interplay between Darwin's theory and various progressive ideologies of the time muddies the water as far as working out causal influences is concerned.

The complex and ever-changing references to Darwin in the social literature suggest that components of his theory had taken on a life of their own in the public imagination, recycled endlessly over successive generations. They had done this independently of the theory's effect on science, which had never been very great. It is as though the theory gained wider credence precisely because it was controversial, not because it dominated the scientific research of the time. Darwinism provided a rhetorical strategy as much as real insights into how the world might work. Many of its components existed independently of the theory—indeed, some were well in place by the time Darwin conceived the idea of natural selection. He even used some of these cultural resources to construct the theory, so it is not surprising that they could be recycled under the controversial label of "Darwinism." One example is the concept of the "struggle for existence," a term actually introduced by Malthus. But the same resources were used to construct other theories in both the natural and social sciences, some of which resembled natural selection in one way or another. This is certainly the case with Spencer's Lamarckian evolutionism. Perceptions of the theory and its associations were then manipulated as new styles of thought emerged. All these factors may have worked to give Darwinism a greater visibility than it actually deserved in terms of its direct impact in the generation of new insights.

Counterfactual history offers us a way of determining whether or not Darwin's key ideas could have had the effects that his critics attribute to them. It is one thing to show that Darwinism was part of the cultural mix that generated certain ideologies and actions, quite another to demonstrate that it was the crucial cause without which none of the attitudes we now deplore could have emerged. If we can make a plausible case, based on what we know about the influence of related ideas, that the same harmful consequences could have emerged and gained scientific justification even if Darwin had not published the theory of natural selection, then that theory can be vindicated from these charges.

This enterprise will not involve an attempt to whitewash Darwinism by suggesting that it is just pure science and that the social consequences were based on misunderstanding or distortion. Darwinism *was* involved, certainly in the promotion of the heartless individualism of the mid-nineteenth-century middle classes, and less directly in the promotion of later, very different models of "progress through struggle." Darwin himself shared some

of the concerns that drove social Darwinism, although he would not endorse the more active efforts proposed to eliminate the unfit.[8] Many other
factors promoted the ideology of struggle and were exploited by Spencer
and later thinkers developing alternative ideas of evolution. If those alternatives could provide foundations for constructing the various forms of
what we call social Darwinism, we shall have to recognize that the theory
of natural selection did not play such a key role. We must also recognize
that both Darwin and Spencer were taken up by ideologues on the Left as
well as the Right. Both evolutionists proclaimed the need for struggle to
drive progress, but both thought that an important result of progress was
the emergence of altruistic instincts. In the race to depict them as the archi-
tects of an evil social Darwinism, their role in promoting policies of social
cooperation has been largely ignored.[9]

To some extent, the very way in which the terminology of the debate
has evolved has managed to shape our perceptions. The terms "neo-
Darwinism" and "neo-Lamarckism" were introduced in the late nineteenth century to denote positions in which natural selection and the inheritance of acquired characteristics respectively were thought to be the
principal mechanisms of evolution. But "Darwinism" had already gained
general currency as the name for evolutionism, especially any evolutionism
involving an element of struggle. This includes Spencer's Lamarckian approach, yet Hawkins and many modern social commentators will not call
Spencer a Lamarckian simply because he included a role for struggle in his
evolutionism. For them, "Lamarckism" means a theory based on willpower
and predetermined goals—more or less what Samuel Butler and the later
neo-Lamarckians intended. So Spencer has to become a social Darwinist
even though he didn't think natural selection was very important! We need
to free our imaginations from these preconceptions and labels, recognizing
that struggle could play a role in Lamarckian thinking too and that the results might look very much like natural selection to the uncritical eye.

Another of Darwin's insights, the theory of sexual selection, provides
a useful model for understanding what was going on. He introduced this
concept in his *Descent of Man* of 1871 to explain human racial diversity,
seeing this as a product of a more general effect through which characteristics are favored because they confer reproductive advantage rather than
survival value. Many commentators link this theory with typical Victorian
attitudes toward the sexes: males are strutting and dominating, females are

coy and choosy. Some imply that Darwin's initiative actually prompted the appearance of these stereotypes in the literature of the time. Yet biologists virtually ignored sexual selection until the mid-twentieth century. The idea that female choice might play a role in evolution was debated—and widely rejected—in other contexts. Women were certainly seen as choosy, but few male commentators were willing to concede any real power to their actions.[10] So the notion that Darwinism *caused* these attitudes toward the sexes to emerge at the time seems implausible. What actually happened was that Darwin built the conventions of his time into his theory, but his contemporaries took those conventions so much for granted that they could not see them as the basis for good science. The situation for natural selection is pretty much the same, the only difference being that here there was a scientific debate, albeit one that led to the theory being sidelined for decades.

Selection and Progress

For all of these reasons, I want to argue that the theory of natural selection could not have had the transforming impact that modern critics claim. Of course, their targets are not always easy to identify. For example, in *From Darwin to Hitler,* Richard Weikart seems to focus specifically on the materialist implications of the Darwinian theory. But in *Hitler's Ethic,* he seems more concerned with the Nazis' hope of breeding an improved human race, and in this respect their ideology can be seen as a product of progressionist evolutionism more than of the selection theory. From the critics' perspective, perhaps it makes little difference which aspect of Darwinism is responsible—the whole evolutionary movement has been a malign influence undermining moral values and the sanctity of human life. So which is the real culprit: the selection theory with its reduction of everything to brute struggle or the idea of progressive evolution with its hope of future perfection? Or doesn't it really matter because both are the agents of Godless materialism? Perhaps we need to think a little more precisely about which form of evolutionism offered the best model for ideologues seeking to justify their positions.

Critics may argue that my effort to show that non-Darwinian evolutionary thinking also had dangerous implications makes too precise a distinction between evolutionism and the theory of natural selection. After all, people in the late nineteenth century were already using the term "Darwinism" to denote the general idea of evolution. If the idea of progressive evolution could provide much of the foundation necessary to justify harsh social policies, then the charge that the theory has been a destructive influence on moral values would still be upheld. But this is a deliberate oversimplification of the situation. The counterfactual approach helps us better understand both the history of evolutionism and the ways in which that history has been manipulated in an attempt to shape modern attitudes toward the theory.

If Darwin's specific explanation of evolution helped to generate a more materialistic vision of the world that would not otherwise have come about, then science had a real and possibly disastrous impact on the development of Western culture. But if non-Darwinian evolutionism could have brought about most of the harmful consequences anyway, the situation becomes much more complex. It is useful for the critics of evolution to sidestep the issue, since by implying that "Darwinism" broadly conceived is responsible they can use the negative image that the theory has acquired to blacken the reputation of the whole movement. The attempt to blame evolutionism as a whole for the ills of the world founders on the implausibility of the claim that the scientific theory actually caused Western culture to put its faith in the idea of progress. All the work of modern historians suggests that the causation runs in the other direction: the growing enthusiasm for progress created the climate of opinion that made it possible for scientists to see the history of life on earth in evolutionary terms. Scientific innovation cannot be solely responsible, even if it contributed to the transformation of the worldview.

Darwin's theory of natural selection was a product of its social environment too, but precisely because it was a somewhat aberrant product, it stands out in a way that makes it easier for critics to claim that it deflected attitudes in a new and dangerous direction. Natural selection was very obviously a new scientific discovery with the potential to make people think differently. But counterfactual history will show that most of the scientific and social development we are familiar with would have followed with or without Darwin, because the trend toward evolutionary thinking was an integral component of social and cultural history in the late nineteenth century. The malign consequences attributed to Darwinism were inherent in underlying trends, not the result of a deflection onto an entirely new path. A reduction in the intensity of the conflict between science and religion would not have meant a bypass of the attitudes stigmatized as social Darwinism.

This point has important consequences for contemporary debates over the role of evolutionism in modern science and culture. The identification of Darwinism with general evolutionism is intended to persuade people, especially those with strong religious beliefs, that modern scientific biology—which is certainly Darwinian to a significant extent—is the heir to a dangerous intellectual trend. Highlighting the allegedly harmful effects of that component from Darwin's time to the present helps to give the impression that scientific evolutionism embodies a materialist perspective intrinsically hostile to religious faith. By showing that the selection theory is not as powerful or as continuous in wider debates as the critics imply, counterfactual history will help us defend science from critics whose arguments depend on the polarization of opinions into black-and-white alternatives. Without natural selection, evolution would have seemed less of a threat to religion—yet its negative consequences would have emerged just as strongly, because those consequences were inherent in wider cultural trends that a single scientific idea could not have deflected.

FREE-ENTERPRISE INDIVIDUALISM

Let's begin with the original version of social Darwinism, often sidelined by modern commentators who identify the term with militarism and racism. Here social Darwinism refers to the extreme form of laissez-faire individualism that flourished in the nineteenth century and came to a head with the robber barons of American capitalism. This was an ideology that seemed Darwinian in the sense that it saw struggle as the natural state of affairs, essential in promoting efficiency and economic progress. In the opinion of its advocates, those who were unable to withstand the pressure to compete deserved little sympathy.

Did Darwin himself endorse such a view of social progress? Few historians now doubt that the ideology of laissez-faire individualism was built into the foundations of his thinking, if only through the influence of Malthus.[11] In the *Descent of Man* he made it plain that selection had acted in human evolution and he worried that modern civilization allows the unfit to reproduce. But, like Spencer, Darwin knew that humans had evolved to live in social groups, and he saw the social instincts thus induced as the foundations of what we call our moral capacity or conscience. Again like Spencer he thought that the social instincts had evolved by Lamarckism, the inherited effects of learned habits, although he also invoked a kind of group selection. Those tribes in which the members cooperated would have an edge over those that were a disorganized mob. Darwin was distinctly uncomfortable with the idea that his theory justified the abrogation of traditional moral values. He was dismayed when a newspaper accused him of proving that "might is right" and of justifying Napoleon and every cheating tradesman.[12] As Thomas Dixon has shown, there were many commentators who recognized that Darwin's real preference was for a society based on social instincts and altruism.[13]

The main source of the aggressive ideology of free enterprise was Spencer—indeed, the classic study of American social Darwinism by Richard Hofstadter calls this episode of history "The Vogue of Spencer."[14] But this immediately raises problems that have been surfacing throughout this study: there were parallels between Darwin's and Spencer's evolutionism, but their theories were not the same, and we cannot simply assume that they made equivalent use of Malthus's concept of the "struggle for existence." Just because Spencer coined the evocative term "survival of the fittest" does not mean that he thought either natural or social evolution proceeded primarily through selection. Nor can we simply assume that the capitalists who praised Spencer were really in tune with the philosophy he was trying to promote—his erstwhile followers could as easily misappropriate him as they could Darwin. Dixon's survey of the rise of altruistic thinking shows that Spencer, like Darwin, hailed the rise of the social instincts and was taken up by socialists as well as capitalists. Because their theories saw social instincts emerging out of an earlier competitive process, they contained resources that both sides in the debate could exploit.

Struggle as the Spur to Progress

At first sight, Spencer certainly looks like a social Darwinist. He was keenly aware of Malthus's principle of population and realized that this would generate a struggle for existence as individuals competed to gain their share of scarce resources. He came close to discovering the idea of natural selection himself and acknowledged Darwin's achievement by coining the term "survival of the fittest." He accepted that struggle was the natural situation in contemporary human society and dismissed the suffering of those who were pushed aside as a temporary evil necessary to achieve a long-term good. It would be only too easy to assume that he must have been imagining that natural selection was the primary agent of both natural and social evolution. But did Spencer really believe that social evolution proceeded by the wholesale elimination of vast numbers of unfit individuals in every generation, which is what an equivalence with natural selection would entail? The answer is certainly no, and it can be backed up by showing that the main role of struggle in Spencer's ideology was to promote a Lamarckian process of self-improvement in the majority of individuals exposed to the pressure of competition.

In discussing the religious implications of evolutionism, we saw that Spencer's philosophy had a deeply moral purpose. He did not want to abrogate traditional morality and envisaged a gradual enhancement of the values associated with the Protestant work ethic: industry, ingenuity, and thrift. Like Darwin, he knew that we had evolved to live in social groups, but, unlike Darwin, he thought that the ultimate goal of evolution was to produce a society of individuals perfectly adapted to their surroundings. Society was like a living organism in which the specialized parts should work together for the benefit of the whole. In the present situation, there was still a tendency for humans to reproduce too rapidly, thus generating the Malthusian struggle for existence. This element of competition was one of the driving forces of mental and hence of moral evolution. When that evolution has reached its goal, struggle will wither away, partly because we shall have learned how to interact more smoothly with our neighbors, but also because as we grow more intelligent, we shall have less energy to devote to reproduction. Spencer believed that thinking and sexual urges drew on the same fund of biological energy, so as the one increased, the other must diminish.[15]

The key difference between Darwin and Spencer, however, was that for the latter the driving force of mental and moral evolution was a Lamarckian process in which learned habits were gradually converted into inherited instincts. Competition (in the present imperfect state of affairs) is beneficial not primarily because it eliminates the unfit, but because it stimulates everyone to become fitter—that is, to adapt more effectively to the social environment. The apparent harshness of Spencer's attitude toward failure conceals the fact that he saw this as a case of short-term suffering working toward a long-term good. Here is Spencer writing in his first important book: "If to be ignorant were as safe as to be wise, no one would become wise. . . . Unpitying as it looks, it is better to let the foolish man suffer the appointed penalty for his foolishness. For the pain, he must bear it as well as he can; for the experience—he must treasure it up, and act more rationally in the future. To others as well as to himself his case will be a warning. And by multiplication of such warnings, there cannot fail to be generated in all men a caution corresponding to the danger to be shunned."[16] Here Spencer is talking about self-improvement at the individual level. Spencer is merely insisting that most people do have the capacity to acquire more effective habits—they are not totally locked in by biological heredity. But when we bear in mind that Spencer was a convinced Lamarckian in his biology, we can see how his appeal to the struggle for existence as a means of stimulating self-improvement encouraged the expectation that beneficial effects would accumulate over many generations to raise the human mind to new levels. Far from being a true social Darwinism, this was a form of social Lamarckism—yet the element of struggle it contains persuaded many of Spencer's readers that it must be a spin-off from Darwin's theory.

Spencer's Influence

Spencer's writings were far more effective than Darwin's in spreading the general gospel of evolutionism. And as one recent survey notes, even informed commentators such as George Eliot frequently blurred the differences between Darwin and Spencer.[17] Later thinkers simply assumed that Spencer's appeals to struggle as the motor of progress must reflect a selectionist perspective. Yet in his later years, Spencer began to doubt the inevitability of progress, partly because the rise of imperialism threatened his prediction that militarism would gradually give way to industrial capitalism. He became far more pessimistic about the short-term future, a change of heart that seems to have escaped the attention of many of his followers. He also became sharper in his attitude toward those who shirked hard work, and in a late book from 1884, *The Man versus the State*, his attack on socialism took a form that seems much more Darwinian, much more concerned about the proliferation of the unfit. But this is an atypical work, and by this time Spencer's reputation was firmly established on the basis of his earlier, more optimistic worldview.

Spencer's philosophy was taken up with the greatest enthusiasm in the United States, where the captains of industry relished his gospel of progress through unrestrained competition. Spencer made a triumphant tour of the country in 1882, culminating in a banquet at Delmonico's restaurant in New York, which was attended by a host of eminent capitalists — but also by many clergymen. Andrew Carnegie became an enthusiastic disciple of his philosophy, along with John D. Rockefeller and railroad magnate James J. Hill. They welcomed his endorsement of competition as the driving force of economic progress, and they all used the vocabulary of the "struggle for existence" and "survival of the fittest" to imply that the process was beneficial because it was natural. They were thus social Darwinists at least in the sense that they used Darwinian rhetoric. Darwin's theory certainly helped to generate the language used to promote the ruthlessness of industrial competition, but was it really an essential model without which the robber barons would have been unable to function?

There are a number of reasons for doubting that a scientific theory, however effective, could have produced such a state of affairs in society at large. Historian Robert Bannister suggested that the popularity of appeals to the "survival of the fittest" has been exaggerated. Many small businessmen feared the rapacious activity of big corporations because they tended to gobble up small fry on their way to dominance over the market.[18] The appeals to Darwinism by those who were succeeding in the struggle often reveal only a superficial understanding of the theory. The Yale economist William Graham Sumner, hailed by Hofstadter as a leading social Darwinist, seems to have been more concerned about the struggle of society as a whole against the limitations of nature. Hofstadter also quotes John D. Rockefeller to the following effect: "The growth of a large business is merely a survival of the fittest. . . . The American Beauty rose can be produced in the splendor and fragrance which bring cheer to its beholder only by sacrificing the early buds which grow up around it. This is not an evil tendency in business. It is the working out of a law of nature and a law of God."[19] The assumption that society must follow the laws of natural evolution is obvious enough, but what has the artificial pruning of a rose tree got to do with the natural selection of random variations in successive genera-tions of reproducing organisms? More seriously, economists even to this day have found it hard to apply the theory of natural selection to their field precisely because there is no obvious analogy between industrial firms and biological organisms. They just don't reproduce in the same way, so there is no room in the analogy for the endless whittling away of variants in suc-cessive generations that the Darwinian mechanism requires. What we have here is rhetoric, not substance. The greed and the ruthlessness of the rob-ber barons were endemic to the society in which they functioned, and it seems unlikely they would have behaved differently had they been unable to use Darwinian language.

The real problem, though, is that these powerful figures were all Spen-cerians, not Darwinians in the modern sense. In terms of personal morality, as opposed to the struggle for economic supremacy, they favored a self-help philosophy that encouraged everyone to make the best use of their talents and abilities. That is why Carnegie founded a string of libraries around the world (all containing copies of Spencer's works). He wanted everyone to have access to improving literature and thus to get the best possible start in life—Spencerians didn't believe in the benefits of formal education. Rock-efeller founded numerous charitable foundations, especially for medical research. Both thought that the accumulation of wealth for its own sake was pointless; the wealthy were supposed to use their profits for the pub-lic good. When Rockefeller argued that the laws of nature were the laws of God, he was echoing the views of the clergymen who also proclaimed themselves Spencerians. This form of social Darwinism was the Protestant work ethic updated to the age of industrial progress. To claim that these at-titudes could not have emerged without the stimulus provided by Darwin's theory is to misunderstand both the attitudes themselves and the relation-ship between ideas and human behavior.

MILITARISM AND NATIONAL CONFLICT

Spencer's later pessimism arose from the confounding of his predictions about social progress by the increasing rivalries between the Western pow-ers. He saw military conflict as a relic of the feudal era and campaigned against the mistreatment of native people in territories colonized by

Europeans. This was the age of imperialism, in which colonial empires came to be seen as the basis of national wealth and prestige. The scramble to open up Africa to exploitation by Europeans only heightened a sense of insecurity, especially among those nations—most obviously Germany—that felt they had been left out. Here was a new form of social Darwinism, although it was actually the first toward which the term was explicitly applied. The struggle for existence was assumed to be between nations and races, not individuals, and the survival of the fittest would establish which culture and political system was the most efficient. Darwinian rhetoric was certainly exploited by the military powers, most vigorously (although not exclusively) by the Germans. It was the prevalence of this way of thinking among the German officer corps that generated the concerns flagged by William Jennings Bryan in his attacks on Darwinism in the 1920s. The same ideology of a conflict to establish the dominance of the superior race and nation subsequently transferred itself into Nazism and sustains the claim that without Darwinism the depredations of Hitler and his followers would not have occurred.

As with free-enterprise social Darwinism, the claim that the scientific theory was somehow responsible for these social consequences can be evaluated at two levels. Is it plausible to believe that the huge geopolitical forces driving the European nations toward war in the period after 1890 could have been deflected onto a less catastrophic track by the absence of what was then considered an outdated scientific theory? It's hard to formulate a detailed response to this question, but I, for one, find it impossible to credit Darwinism with having so great an influence on the global rivalries that led to war. European nations were gearing up for confrontation for a host of economic, social, and political reasons and would hardly have been deterred by the absence of a scientific model of conflict. The same point holds for the events leading to World War II. It was the disaster of the Great War that created a social environment favorable to the rise of an extremist party such as the National Socialists (Nazis). They too were spoiling for a fight and were inclined to find a scapegoat for the national humiliation suffered in the previous conflict. For both of these episodes, it makes sense to argue about the extent to which Darwinism might have shaped the rhetoric of national hostility, but it is hardly plausible to blame it for the conflicts themselves. Whether Darwinism could have helped generate the specifically racist aspects of Nazi ideology is another question addressed later in this chapter.

I must concede that Darwinism did become involved with the culture of imperialism, providing a source of extremely effective rhetoric. The links exploited by this Darwinian language were also perfectly genuine. Although Darwin's key insight was the theory of natural selection acting at the individual level, he did make use of the idea of group selection in ways that would permit an analogy with human tribal and national conflict. But admitting that his theory became involved with the ideology of imperialism is not the same as admitting that it was a catalyst of warlike rhetoric and attitudes. As with individualism, Darwin's theory was integrated with wider scientific and social movements, and ideas that are part of this broader dimension also existed independently of the core theory. Add to this the fact that the theory was used in a multitude of contradictory ways. The most thorough study by a modern historian of militaristic social Darwinism points out that there were a number of peace activists who made use of the theory to bolster their arguments.[20]

Why should the emphasis have switched from individualistic social Darwinism to the nationalistic form in the last decade of the nineteenth century? The two ideologies are not only different—they are mutually incompatible. Those who wanted a strong state to guard against threats from rivals distrusted uncontrolled capitalism (after all, an arms manufacturer who sought only profit might sell to a rival power). The Germanic worship of the state almost certainly owed more to the idealist tradition in philosophy than to the individualism that underlies Darwin's thinking. The assumption that struggle is endemic to the relationship between groups was widespread long before Darwin published and was later accepted even by writers who openly rejected the theory of natural selection.

All these factors suggest that the change of focus from individualism to imperialism was driven by wider cultural and social factors, each of the two phases seeking whatever support they could find from science. Counterfactual history will help us to judge the real impact of Darwinism by allowing us to see how these other forces might have substituted for Darwin's theory when it came to seeking justifications for the rise of imperialism. If we can show that these other factors could have fit the bill, then we refute the claim that Darwinism was a necessary feature of the most warlike rhetoric.

There are two areas where we can actually see an element of group struggle in Darwin's own thinking. The first is his appeal to tribal conflict as a means of explaining the origin of social instincts in the earliest humans. Far from being a direct product of his theory of natural selection, this later addition drew upon resources available to anyone else at the time. The second area provides a less direct link between animal and human evolution. Darwin elaborated an entire worldview on the basis of his theory of natural selection, the most important component of which was the notion of branching, divergent evolution driven by adaptation to new environments. He appealed to biogeography to show how populations became divided by migration to new locations, followed by subsequent adaptation. It was an integral component of this theory that the invading population often displaced an indigenous species and drove it to extinction. The language that biogeographers used throughout the nineteenth century is replete with the metaphors of invasion and conquest. The link with Darwinism is plain— but the link is with the biogeographical perspective, not with the idea of natural selection.

Tribal Conflict

In his *Descent of Man*, Darwin endorsed the view that competition between individuals had promoted the development of human intelligence. But he faced a real problem explaining how this level of struggle could generate the cooperative, altruistic instincts that he believed were the foundation of our moral sense or conscience. Surely a self-sacrificing individual would quickly be eliminated in the struggle for existence? Darwin agreed with Spencer that the inherited effects of learned habits would be important, but he looked for something more. It was at this point that he invoked the possibility that selection acting between family groups or tribes might have played a role, and it was the only point at which he deviated from his normal reliance on individualism. (Most modern biologists think that individual selection is dominant, although there has been a revival of interest in group selection in recent decades.) He argued that tribes in which the members shared cooperative instincts would tend to be victorious in the struggle for resources over those less cohesive: "At all times throughout the world tribes have supplanted other tribes; and as morality is one important element in their success, the standard of morality and the number of well-endowed men will thus everywhere tend to rise and increase."[21] Later writers might well have seized upon this passage to justify national competition (although very few actually read the *Descent of Man*). But note that Darwin's purpose was not to promote warlike instincts or slavish devotion to the state—on the contrary, it was to explain the origin of personal morality.

Darwin referred to articles written by Walter Bagehot, subsequently reprinted as his *Physics and Politics* of 1872. Bagehot's interest in natural selection was confined to the level of group conflict, and he seems to have had little interest in biology. His main concern was to argue that in the ceaseless struggle between peoples that had plagued most of human history, anything that enhanced loyalty to the group, including religion, had been an advantage. There is thus every reason to believe that this aspect of Darwin's thinking, far from being an extension of his core theory, was a later addition prompted by an external source. He might have noticed the role of group conflict earlier—after all, Malthus had introduced the term "struggle for existence" in the context of warlike tribes, and the sixth edition of his book (which Darwin read) had an extended description of the wars of extermination that have taken place among savage cultures. Darwin had initially translated this into the idea of individual struggle to create his theory, and, thanks to Baeghot, he only now came back to the concept of group struggle.

Malthus's section on the role of tribal struggle reminds us that the idea was widely available throughout the nineteenth century. The Victorians knew their Bible, and the Old Testament contains several passages in which the extermination of whole peoples is described and even endorsed. Victorians were also educated in the classics and would have been familiar with the destruction of Troy, Rome's war of annihilation against Carthage, and Caesar's conquest and enslavement of the Gauls. The war against Napoleon was still fresh in everyone's mind, and to drive the message home, there was the rebellion against the British in India in 1857 (which the British called a "mutiny" and put down with great ferocity) and the American Civil War. Spencer and his followers were appalled at these lapses from the path of industrial progress, but toward the end of the century there was a growing feeling that such national conflicts were inevitable and were, perhaps, the best way of sorting out which culture was the most effective. Few of those who went down this route felt much sympathy for the ideology of free-enterprise individualism within which Darwin's and Spencer's ideas were conceived. It is hard to believe that in a world without Darwin, they would have been deterred by the lack of the individualistic theory of natural selection.

In our world, the imperialists certainly used Darwinian terminology. And in a few cases they were genuine scientific Darwinists. The best example here is Karl Pearson, one of the founders of the modern statistical approach to natural selection, who railed against the inefficiency of the British Empire as revealed by its initial failures in the Boer War in South Africa.[22] Pearson simply abandoned natural selection within the population in favor of eugenics, and he translated the struggle for existence into nationalistic terms so that he could appeal for a strong centralized state. It may be no coincidence that Pearson was deeply immersed in German culture, because it was certainly in that country that the ideology of a militaristic state imposing its will on the world by brute force achieved its most sinister level of activity in the period leading up to the Great War. There were sound geopolitical reasons for the Germanic desire for self-assertion: Germany was surrounded by strong rivals (Britain and France to the west, Russia to the east) and also felt that its late emergence as a unified state had excluded it from the race to acquire colonies. But to what extent did Darwinism inspire its militarism, and what other conceptual resources was it able to draw upon?

The late nineteenth century certainly saw the rise of what has been called "conflict sociology," especially in Germany, although not all it its exponents made explicit appeals to Darwinism. The most widely cited example of the use of Darwinian language came in General Friedrich von Bernhardi's *Germany and the Next War* of 1912. Bernhardi saw the relationship between nations as a global struggle for existence and asserted that this arena was governed solely by the principle of "might is right." But several modern commentators have argued that the idea of natural selection played a relatively subordinate role in his thinking, especially in comparison to the desire to exalt the power of the German state and the kind of realism that one would expect from a military man (although in fact Bernhardi was regarded as a bit of a loose cannon by the German high command).[23] Bernhardi also knew his history. He quoted Heraclitus to the effect that "War is the father of all things" and observed: "The sages of antiquity long before Darwin recognized this"—hardly a comment one would expect from someone whose chief inspiration was the scientific theory.[24]

Perhaps Darwin encouraged people to think that there was no progress or purpose in a world based on brute force. But that was precisely not what the advocates of German supremacy intended—they were convinced that struggle was the means by which the highest form of culture demonstrated its superiority over its outdated rivals. Far from being an extension of the Darwinian view of history, this was a continuation of the traditional conservative view that the trajectory of human history can be seen in the successive rise, fall, and replacement of dominant powers. The new Germany was the belated triumph of the Teutonic culture that had emerged as the successor to the empires of Greece and Rome. Bernhardi himself included a chapter in his book on Germany's historical mission to civilize the world.[25]

A crucial component of the imperialist ideology was another profoundly un-Darwinian way of thinking, the subordination of the individual to the state and its leaders. This had its origins in the idealist philosophy of Hegel, which rejected the free-enterprise system in favor of the view that an individual's life was only properly expressed as part of his or her surrounding culture. Heinrich von Treitschke and Leopold von Ranke encapsulated this vision of German history long before Darwinism became fashionable. Their philosophy was translated into what became known as "conflict sociology" by Ludwig Gumplowicz, but again without the direct influence of Darwinism. If there was a biological input into this way of thinking, it came not from evolutionism but from an analogy with the organism and its component cells, all of which collaborate to ensure the successful functioning of the whole.[26]

The resulting political system stressed the role of state power and increasingly began to exalt the leader of the state as a symbol of the culture's mystical unity and purpose. With Arthur Schopenhauer and Friedrich Nietzsche promoting the philosophy of the will, the ideology of power and dominance became even more tightly focused on the leader—with obvious implications for the future emergence of Nazism. It is hard to imagine a way of thought more remote from the utilitarian individualism of Darwin and Spencer than Nietzsche's philosophy of the "superman" whose willpower is the sole source of inspiration.[27] In a world without Darwin, the ideology of state power would have developed under its own resources, and it might even have been more effective than it was in our own world.

Dispersal and Displacement

The other area where we can see an element of group competition in the Darwinian worldview is in its use of biogeography to explain the historical relationship between species. Darwin was led to his theory of natural selection because he was already convinced that new species form when members of an original population become separated by geographical barriers and adapt to new environments. Migration to new territories was the key to divergence, and Darwin's growing sense of the harshness of nature alerted him to the possibility that invading forms often gained a foothold only by displacing an area's original inhabitants, usually resulting in their extinction. The process had obvious parallels with the way in which European powers were conquering and colonizing other parts of the world, and the voyage of the *Beagle* was, of course, part of Britain's effort to explore and dominate the wider world.

Biogeography provides an obvious vehicle by which Darwinian ideas could be conveyed into the ideology of imperialism. But the situation is more complicated than critics of Darwinism might imply, because the biogeographical model of dispersal and displacement emerged independently of the theory of natural selection. The two ideas certainly fit together well, but dispersalist biogeography did not emerge from the selection theory. Indeed, Darwin had formulated his vision of divergent evolution before conceiving natural selection. A key inspiration here was Lyell's geology, which helped to convince a number of naturalists that extinction was a normal and inevitable response when species were confronted with an ever-changing environment. The image of nature as a scene of constant warfare between species (as well as individuals) was becoming widespread in the middle of the century, some time before Darwin published. The biogeographical perspective would have existed even without Darwinism, although no doubt it gained more prominence because of its popular association with the selection theory. In the non-Darwinian world, there would have been no shortage of scientific metaphors available to the imperialists.

The metaphor of the "war of nature" had been widespread even in the late eighteenth century among writers such as Erasmus Darwin, although at this point it was assumed that its end product was ultimately beneficial. Darwinism was part of the process that undermined faith in natural theology, but it was by no means the only factor involved. Geologists convinced everyone of the reality of extinction, and Charles Lyell's uniformitarian vision of history made it clear that the death of species was due not to isolated catastrophes but to the everyday vicissitudes of natural change. Lyell and the French botanist Alphonse de Candolle both stressed the endless competition between species. Tennyson's *In Memoriam* not only depicted nature as "red in tooth and claw," but it also stressed her indifference to the fate of species: "A thousand types are gone: / I care for nothing, all shall go."[28] Darwin and Wallace were by no means the only naturalists to realize that within such a worldview, the dispersal of species was almost certainly going to result in the extinction of any less well-evolved competitors in the territories they invaded. Alfred Newton, a Cambridge university professor of zoology, noted the growing tendency for human activity to threaten the very existence of some species, and everyone knew about the end of the dodo in Mauritius. German naturalists, too, were well aware of the dangers of extinction.[29]

Wallace's *Geographical Distribution of Animals* of 1876 initiated the main wave of enthusiasm for dispersalist biogeography, and Wallace would probably have undertaken this project even if he were not working within a Darwinian framework. Wallace was a socialist and no enthusiast of imperialism, yet even he slipped unconsciously into the language of invasion and colonization when describing animal migrations. Metaphors of imperialism were woven into the language of biogeographers throughout the late nineteenth century.[30] Yet many of these scientists were not enthusiasts for the theory of natural selection, and some were outright opponents. The idea that selection acted as a negative process for weeding out the less successful of evolution's products flourished even among anti-Darwinian naturalists who dismissed any thought of selection actually producing new species. There is thus no reason to assume that the metaphors of imperialism would have been muted in the biology of a world in which Darwin had never published.

This point can be demonstrated in the area where dispersalism came closest to politics: theories of human origins. Until the 1890s, most palaeoanthropologists assumed that humans had evolved gradually from an ape ancestry, and most saw the brutish Neanderthals as an early stage in this process. But around 1900, there was an abrupt change of emphasis as students of fossil hominids began to argue that the ancient types known from the fossil record were not our ancestors. They were parallel branches of our family tree only distantly related to modern humanity. It suddenly became fashionable to think of the Neanderthals as remnants of one of these earlier types, abruptly wiped out when modern humans invaded Europe in the late Paleolithic era. Two of the leading advocates of this view were Arthur Keith and William Johnson Sollas, and both were only too keen to point out its lessons for the current state of affairs. Keith developed a whole theory of evolution based on racial competition, while Sollas asserted that the evolutionary process expressed the philosophy of "might is right," and he insisted that any race that did not maintain its alertness would incur a penalty that "Natural Selection, the stern but beneficent tyrant of the organic world, will assuredly exact, and that speedily, to the full."[31]

This surely looks like an outgrowth of Darwinism, and Keith at least always called himself a Darwinian. Yet he showed little understanding of individualistic natural selection and thought that variations were directed along purposeful channels by the action of hormones within the body. Sollas openly ridiculed the view that natural selection could actually create new species, dismissing it as "an idol of the Victorian era."[32] Clearly, their vision of competition was confined to its action at the group level—there was no appreciation that Darwin's theory had solved the problem of how new species were actually produced. This whole episode in paleoanthropology, with all its consequences for reinforcing the imperialist ideology so prevalent at the time, could have unfolded even if the Darwinian theory had not been in play.

RACISM

Keith postulated conflict between groups defined by their racial origin, and this moves us on to the contentious topic of the link between evolutionism and the attitudes we now associate with racism. There is no doubt that the evolutionary movement became involved with Europeans' efforts to define nonwhite races as biologically inferior. This was true everywhere, but nowhere more than in Germany, where some commentators have linked Ernst Haeckel's Darwinism to the origins of Nazism. But once again we encounter the point that "becoming involved with" is not the same as "causing," and on this topic even Weikart concedes that the idea of a hierarchy of races existed long before Darwin published.

Evolutionism became entwined because it offered a plausible explanation of why some races might not have advanced as far as others up the scale leading from the ancestral ape. Most of the Darwinians endorsed this way of thinking (with the notable exception of Wallace). But they remained resolutely opposed to the most extreme form of race science, which treated the races as distinct species. Non-Darwinian evolutionists who thought that several different lines of evolution had given rise to human types promoted this form of race science far more strongly. In the twentieth century, the rise of modern Darwinism coincided with efforts to demonstrate the genetic unity of the human species against this extreme model of racial difference. We can make a plausible claim that in a world without Darwin, race science would have been even more vigorous and racist attitudes in society at large more powerful.

The Racial Hierarchy

The perception that nonwhite races were inferior to Europeans had developed over the centuries, deriving much of its support from the growth of the slave trade. It suited slave owners to believe that the people they misused were not their mental or moral equals. Even in the eighteenth century, anatomists argued that the black races had receding foreheads and smaller brains. They were even depicted as having apelike features, long before anyone had begun to imagine that humans were descended from apes. This was an expression of the old "chain of being" in which all species were linked into a single, linear hierarchy with humans at the top. The German anatomist J. F. Blumenbach, a founder of the structuralist tradition in biology, had a world-famous collection of skulls upon which such generalizations about the races were based. The middle decades of the nineteenth century saw a renewed interest in the physical anthropology of race by anatomists such as Robert Knox. There was particular emphasis on brain size as an indication of intelligence—a legacy of the phrenological view that the brain is the organ of the mind. Paul Broca in Paris and James Hunt in London founded societies dedicated to showing the degree of anatomical difference between the races, with particular emphasis on skull dimensions. They produced masses of evidence (now largely dismissed as spurious) purporting

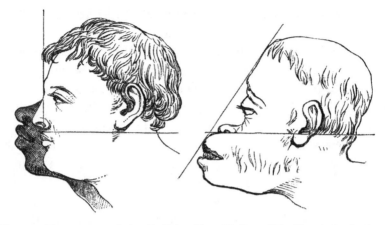

Figure 9 · Human races as depicted by Robert Knox, *The Races of Men*. Note the implication of apelike features in people of African descent, even though Knox was not an evolutionist.

to establish a hierarchy of racial types based on brain capacity and other features.[33] Neither had any interest in showing the evolutionary origin of humans from apes.

At the same time, cultural anthropologists such as Edward B. Tylor and Lewis H. Morgan began to formalize a hierarchical arrangement of human cultures. They placed hunter-gatherers at the bottom of the scale, with agriculturalists ranking above them, and finally commercial and industrial civilizations such as modern Europe at the top. This was conceived from the start as a historical sequence: modern "savages" were actually primitives—relics of the earliest stages of human society and culture preserved in backwaters isolated from the mainstream of progress. This model articulated well with the discoveries of prehistoric archaeologists, who had finally established clear evidence that Stone Age cultures had existed on earth long before recorded history began. Historians agree that this progressionist model of human history (and its application to modern "primitives") emerged coincidentally with but independently of the Darwinian revolution in biology.[34]

Some cultural evolutionists—most notably Tylor—did not at first believe that people with the lowest level of culture were intellectually inferior to Europeans. But the two movements soon came together as Darwinists such as Sir John Lubbock began to interpret our Stone Age forebears as ancestral types that could be expected to show evidence of their recent emergence from the apes. In the absence (as yet) of fossil hominids, modern savages were treated as equivalent to these primitive ancestors, and the physical anthropologists' alleged evidence of small brains and apelike features in the "lowest" races was called in to confirm the link. Darwin certainly contributed to this process in his *Descent of Man*, and in Germany Ernst Haeckel built the idea that the human races show different levels of development firmly into his Darwinism. The recapitulation theory was called in to imply that the lower races were immature versions of the highest form of humanity, presumed to be the Europeans. Their development was frozen at an earlier stage before the appearance of the higher mental powers. The criminal anthropologist Cesare Lombroso even argued that criminals and other degenerates were throwbacks to an earlier stage of human evolution.

All of these representations of racial hierarchy projected the image of a linear hierarchy of evolutionary stages. Darwin himself had initially shared the view that all humans were biologically equivalent, their cultural variation owing to the different conditions under which they lived. His branching model of evolution did not fit very well with the idea of a simple hierarchy of cultural or racial types. The evidence suggests that the progressionist vision of human history as an ascent toward European civilization and brainpower would have been created whether or not the *Origin of Species* was published. Far from being absent in the world without Darwin, that linear model would have been even more powerful because it would have faced a less effective challenge from the theory of divergent evolution.

The Typology of Race

For many nineteenth-century thinkers, the really crucial question was how closely the various human races were related. Were the cultural (and possibly intellectual) differences between them merely local variations within a single human type, all of whom shared a common origin? This was the position known as "monogenism," and in its original form this meant that all humans were descended from Adam and Eve (evolutionists had to modify this to imply descent from a single ancestral population). But those who were increasingly conscious of racial diversity suspected that the differences were far too great to have emerged in the few thousand years provided by the biblical timescale. This might have seemed less of a problem once a proper awareness of human prehistory emerged in the 1860s, but the physical anthropologists who tried to establish the identity of the main racial types insisted that they could not possibly have evolved from a common ancestor. They were distinct species with separate origins, the position known as "polygenism."

Polygenists could not believe that all humans had descended from Adam and Eve, but there was an old, if rather heretical, theory that held that the nonwhite races had evolved from separately created "preadamite" ancestors.[35] In the early nineteenth century, polygenist ideas were widespread, largely inspired by what we have called the structuralist movement in biology. The possibility that each region of the world had somehow generated its own form of humanity was taken quite seriously—which is why physical anthropologists such as Broca and Hunt could sidestep the notion of a single origin. The Swiss American biologist Louis Agassiz explicitly extended his extreme form of what we would call creationism to include separate miracles for the origin of each race.

Far from originating with Darwinism, non-evolutionary systems sustained this extreme interpretation of racial diversity until the 1860s. Darwin himself came from a family passionately opposed to slavery, and he was appalled by the cruel treatment he saw meted out to slaves in South America on the *Beagle* voyage. He was a monogenist, committed to the view that all the races share a common origin and a common humanity. A recent study by Adrian Desmond and James Moore argues that this model of geographically separated races having diverged from a single ancestral population provided the inspiration for the more general vision of divergent evolution that was the prelude to his discovery of natural selection.[36] Later in the century, Darwin and his followers refused to endorse the racist physical anthropology that Broca and Hunt peddled. Although Darwin and company conceded that some races had evolved higher levels of mental and moral powers than others, they refused to believe that the differences were sufficient to count them as distinct species. After all, they could interbreed, a point the polygenists preferred to ignore. Wallace refused to endorse even the most limited form of racism—his experiences with the native peoples of South America and the Far East convinced him that all races had more or less the same mental powers.

The evolutionary movement of the late nineteenth century did not, however, eliminate polygenism altogether. Some of the non-Darwinian theories that flourished even in our own world subverted the logic of common descent by suggesting that inbuilt variation trends could drive parallel lines of evolution toward almost the same goal. It was relatively easy to apply this model to the human races, arguing that any link lay far back in the past, long before they emerged as humans. The similarities that Darwinians interpreted as evidence of a recent common ancestor were thought to have developed independently in separate species, each driven by the same trend. It was then even easier to suppose that some species had evolved further up the scale of mental development than others. In its most extreme form, as proposed by Karl Vogt, this model supposed that each race had its origin in a separate ape species, whites from chimpanzees, blacks from gorillas, and so on. Although few authorities went this far, the theories of human origins proposed by paleoanthropologists such as Keith and Sollas all postulated an extreme antiquity for the different racial types. Parallelism also figured strongly in the thinking of fossil-hunter Louis Leakey and in the orthogenetic vision of human origins promoted by Henry Fairfield Osborn. Mid-twentieth-century Darwinists such as Julian Huxley played a major role in combating these racist views, especially once their consequences became obvious in Nazi Germany.[37]

This brings us to the vexed question of the role played by German Darwinism in the creation of Nazi ideology. There can be no doubt that German evolutionists, most obviously Ernst Haeckel, were active in promoting the idea that some races were more highly advanced than others. There is much debate over the extent to which Haeckel was anti-Semitic, Daniel Gasman in particular arguing that this was a key feature of his monist worldview and a major contribution to Nazi thinking.[38] I am no expert on German culture and will pass no judgment on this topic here, but it is in any case a side issue to the main point I wish to make. I believe that the extent to which Haeckel actually used Darwin's theory of natural selection was very limited. His progressionist evolutionism, including the component of racial differentiation, would have emerged in the late nineteenth century whether or not Darwin had published the *Origin of Species*. It just wouldn't have been called "Darwinism," and the concept of selection would have had a more limited role in it. In a world without Darwin, theories of directed evolutionism and parallelism would have been even more influential, so it seems reasonable to imagine that the extreme form of scientific racism would also be more powerful. The idea that the races are separate

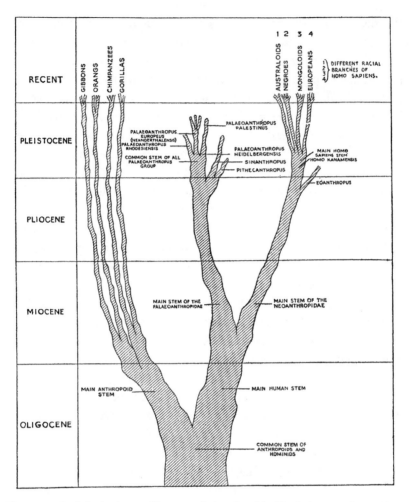

Figure 10 · Louis Leakey's tree of human evolution, from his *Adam's Ancestors* of 1924. Note that the Neanderthals and other extinct relatives of humanity form a parallel line of development independent of the one leading to modern humans, while the living races trace their ancestry back to the early Pleistocene (which in modern dating would make them nearly two million years old). Leakey subsequently made important discoveries of fossil hominids, and his later writings placed less emphasis on parallelism.

species with long independent histories resonates easily with the mystical German notions of a *Volk* with its own inbuilt cultural values. Hitler was not the only architect of Nazi ideology, and Heinrich Himmler seems to have believed that the Aryan race derived from forebears with supernatural powers.

Counterfactual history helps us to see that far from being the cause of scientific racism, Darwin's theory might actually have helped to limit its influence. Without trying to pretend that Darwin and his British followers were completely free of racial prejudice, we can see that they resisted the more extreme theories of racial difference because they did not accept the idea that built-in trends could independently create a group of species capable of interbreeding. This was exactly the point made by later Darwinians such as Julian Huxley, and it has become the cornerstone of modern thinking on the issue. The true sources of polygenism were the creationism of the preadamite theorists, the structuralism of the early nineteenth-century Continental biologists, and the nonselectionist evolution theories of the period around 1900. All of these influences would have been more active in a world without Darwin, and they would have provided the ideologues with a more clearly articulated theory of racial inequality.

EUGENICS

Our last topic is a movement that is often counted as a form of social Darwinism, although it arose from a growing conviction that natural selection was no longer effective in maintaining the quality of the human race. It was Darwin's cousin, Francis Galton, who began to argue that individual character is rigidly determined by heredity. In the 1860s and 1870s he was largely ignored, since his views contradicted the Spencerian vision of social progress though individual self-improvement. According to Galton, people couldn't improve themselves beyond the capacities they inherited, and this meant that selection was the only way of changing the overall character of the population. Yet in a civilized society, we do not restrict the ability of people to have children, which means that even those with the lowest mental and moral capacities continue to breed. Both Darwin and Galton worried that this might lead to degeneration, and toward the end of the century there were growing fears that this was indeed the case. Modern industrial society had created slums in the great cities where the least fit members of the human race congregated and reproduced without restriction. The professional classes meanwhile were restricting the size of their families. The consequence, warned Galton, would be a steady degeneration of the race as the proportion of unfit individuals increased.[39]

Galton's solution was his eugenic program, in effect a call to impose a mechanism of artificial selection on the human race. The fittest were to be encouraged to have more children, while the least fit would be discouraged or even prevented from breeding. By the turn of the century, the middle-classes became increasingly willing to accept the claim that character is predetermined by heredity, and the eugenics movement gained wider support. It was active in many countries throughout the early decades of the twentieth century. In Britain there were moves to detain the "feeble-minded" in institutions where they would be segregated by sex. But there were increasingly active calls for the sterilization of the least fit members of society, and sterilization programs were introduced in a number of American states. This movement became particularly active in Germany, where the Nazis then built on it in their campaign to "purify" the Aryan race. In the end the Nazis went beyond mere sterilization and began to exterminate those elements of society they wished to suppress. This included not only the Jews and other racial types, but also anyone whose behavior did not fit their desired pattern.

To the critics of Darwinism it seems obvious that eugenics was a product of the theory they distrust. The link between Darwin and Galton is clear, even to the level of the family connection, and Galton's disciple, Karl Pearson, was both a eugenist and a founder of the statistical approach to population studies that reformed the theory of natural selection in the early twentieth century. Hitler's murderous policies seem only to reintroduce the element of ruthlessness embodied in the vision of nature as a scene of endless slaughter. But as with all the other forms of social Darwinism, closer inspection shows that the link between science and ideology is too complex to sustain the charge that Darwin's theory actually caused the emergence of eugenic policies. The selection theory was certainly involved, because many thought that eugenics was essential to replace natural selection as a means of winnowing out the unfit and preventing degeneration. But other scientific factors were involved too, and the growing conviction that heredity determines character seems to have emerged more from changing social attitudes than from the science itself.

I argued in chapter 6 that we can explain how both eugenics and a new science of heredity could have emerged in a world without Darwin. Heredity became a matter of social concern because the middle classes began to fear that the unfit members of the population were outbreeding them.

It was this new social attitude that began to focus biologists' attention on the question of heredity, leading in 1900 to the rediscovery of Mendelism and the emergence of genetics. This in turn undermined the credibility of Lamarckism and led to the reinvigoration of Darwinism. But in a world where Darwin had not already appealed to the model of artificial selection to explain natural change, this would be the point at which that connection would be made and the theory of natural selection conceived for the first time. It was eugenics that encouraged scientists to focus on heredity and recognize the potential of artificial selection, and they could have done this without the inspiration of Darwinism.

Heredity and Politics

Francis Galton was anxious to make his mark in the scientific community and decided to focus on heredity as an area where he might make an impact. His book *Hereditary Genius* of 1869 tried to show that clever people had clever parents and implied that lower levels of intelligence would also be inherited. He became increasingly concerned that the less able were breeding more rapidly than the professional classes (whom he assumed to be of superior intellect) with the result that the biological quality of the population was decreasing. Eugenics was his remedy, and he increasingly presented it with almost evangelical fervor as a means of restoring the nation's ability to command its empire.

To begin with, no one paid much attention. Even Darwin, for all that he worried about the worst individuals still being allowed to breed, did not think that people varied much in intelligence. Like Spencer, he thought individual differences were more a case of varying industry and application, so that everyone could be induced to improve themselves if suitably stimulated. In the 1860s and 1870s it was this philosophy of self-improvement that dominated political thought, creating the first wave of social Darwinism. Only in the last decades of the century did doubts about the malleability of individual character begin to flourish. The anticipated improvements in the quality of the human race did not seem to have come about—on the contrary, there were now fears that the race was degenerating as an ever-increasing army of the idle and the stupid bred like rabbits in the slums of the great cities. The middle classes did not want to see their hard-earned salaries eroded by taxes that would be thrown away on those who could not, they now argued, be expected to benefit from improved conditions and better education.

People had always recognized that some aspects of character might be hereditary—the taint of insanity was something to be hidden away in the history of any respectable family.[40] In the era of Spencer this was seen as an exception, but now that his recipe for social evolution seemed to have failed, it became politically expedient in some quarters to insist that heredity predetermines every aspect of character. No amount of environmental change could be expected to produce social improvement, and the only way of preventing degeneration was to artificially restrict the breeding of the least fit members of society. The feebleminded should be discouraged or even prevented from having children, either by institutionalization or by actual sterilization. It would also help, of course, if the professional classes could be persuaded to have larger families. In effect, the state would introduce a policy of artificial selection to shape the biological quality of its population.

It is hard to see this new ideology being inspired purely by biological theory. After all, Darwinism was now in eclipse, and there had been little agreement over the mechanism of heredity. It was increased social interest that actually seems to have focused biologists' attention on the question, leading both to Galton's work and to Weismann's concept of the "germ plasm" as the material transmitter of hereditary characteristics. Eventually, the rediscovery of Mendel's laws created the model of individual unit characteristics being transmitted unchanged from one generation to the next, unaffected by the environment in which they operated.

There were now calls to restrict the transmission of what were seen as harmful genetic units in the human race, and it was inevitable that interest would begin to focus on the method of artificial selection used by animal breeders. For centuries, breeders had worked toward the improvement of cattle, sheep, dogs, and pigeons, convinced by practical experience that heredity was indeed the key and that to improve the breed, one had to select the best individuals and allow only them to reproduce. Yet academic biologists had largely ignored their work until Darwin took an interest in his search for a model on which to base a theory of natural evolution. Darwin's work was innovative precisely because no one else in the scientific community of the early and mid-nineteenth century had been prepared to see the breeder's methods as a guide. Without him, the selection model would have been ignored for several more decades. But now, in the decades around 1900, the time was ripe. Eugenics demanded a theory of hard heredity, a theory in which characteristics are determined by heredity and not by the environment. But it also demanded a role for some form of selection, which turned attention to the breeders' methods.

Heredity and Selection

In our world, Darwin's theory was caught up in the emergence of the eugenics movement, but its involvement was far from straightforward. Apart from the delay of several decades between the original Darwinian debate and the rise of interest in the social implications of heredity, there was a degree of looseness between the thinking of Darwin and many of the exponents of hereditary determinism. One can believe that heredity determines character without committing oneself to the idea that natural selection is the process that determines the genetic makeup of populations. Far from being Darwinists, Galton and many of the early geneticists contributed to the process that brought the selection theory into eclipse around 1900. It is thus perfectly possible to imagine a counterfactual world in which there was no *Origin of Species*, but the eugenics movement emerged more or less as we know it.

Reading the *Origin* clearly inspired Galton to begin his work on heredity, but it was Darwin's appeal to the analogy with artificial selection that caught his attention, not the theory of natural selection itself. When Galton discussed the process of species formation in his *Natural Inheritance* of 1889, he made it quite clear that he did not believe that the natural selection of ordinary individual differences was enough to transform a population into something that would count as a new species. Selection could modify the species within certain well-defined boundaries, but when its effect was removed, the population would revert to the previous norm. To create a new species required a saltation, a substantial new characteristic introduced quite suddenly by a process going beyond normal variation. There was little suggestion in Galton's program that a eugenic policy could produce a new and improved form of humanity. At best, artificial selection could prevent degeneration and perhaps elevate the population to the higher end of its range of normal variation. The presumed advantages of applying artificial selection to the human race did not rest on the assumption that this would mimic the process of natural evolution.

It was Karl Pearson who showed that Galton's understanding of the laws of heredity was flawed and that natural selection acting on normal variations could produce a permanent effect on a species. Pearson was also an imperialist who believed that in human affairs, the focus of struggle had shifted from the individual to the state. He endorsed Galton's eugenic policies as a means by which the state could control the population and maintain its biological qualities. There was certainly a widespread assumption that the relaxation of natural selection in human populations would lead to degeneration. This sustained many calls for the introduction of artificial limits on the reproduction of those deemed unfit. A detailed study has shown, however, that Pearson kept his work on natural selection in wild populations quite separate from his studies of human heredity.[41] The two programs used totally different methodologies. Even here, then, the link between Darwinism and the eugenic program was muted, and there were many who wanted to institutionalize or sterilize the unfit who saw no parallel between this and Darwin's explanation of the process of natural evolution.

In Germany, eugenics also became closely associated with the politics of race. The British eugenists were certainly aware of hereditary differences between the races but paid little attention to the issue. But in both America and Germany the perceived threat posed by "inferior" races outbreeding the Anglo-Saxon elite was central to eugenic concerns. Race mixing was widely condemned—the German term for eugenics translates as "race hygiene." Hitler's eugenic policies were explicitly intended to preserve and even improve the Aryan race. Such concerns for racial purity were—as we have already seen—a product of non-Darwinian evolutionary ideas, not of the theory of natural selection. The Germanic approach to Darwinism was associated more with the idealist vision of the state as the embodiment of a nation's cultural identity than with the individualism that had inspired Darwin and Spencer.

We can confirm the weakness of the link between eugenics and Darwinism by looking at the movement that in our world transformed the study of heredity in the early twentieth century: genetics. There was certainly a strong link between genetics and eugenics, especially in America and Germany. The idea that there were discrete, fixed hereditary character-

SOCIAL EVOLUTIONISM · 269

istics circulating in the population was widely used to support the claim that some of these genetic units produced feeblemindedness and other defects. Preventing the reproduction of those who carried these faulty genes would soon eliminate them from the population. Yet the majority of the early geneticists were not Darwinians. Like Galton, they believed that large mutations (saltations) would be needed to produce a new species; natural selection acting on the normal range of variation was not enough. The geneticists' support for eugenics was based on their work in animal and plant breeding. They saw its presumed effectiveness in shaping the human population as the equivalent of artificial selection in the breeding of domesticated species. They rejected the Darwinian theory of natural selection, which is why Karl Pearson was suspicious of their new model of heredity. As a result, it took several decades before Pearson's work on selection in wild populations could be synthesized with genetics to produce the modern theory of natural selection.

These facts are difficult to reconcile with the claim that eugenics was a belated product of Darwin's thinking on evolution. The social developments that gave rise to the eugenics movement focused scientists' attention on the study of heredity, but the resulting theoretical innovations were not driven by, and in some cases were actively contrary to, the theory of natural selection. What the eugenists wanted was a theory of hard heredity and a justification for applying a process of artificial selection to the human race. They could find this quite easily in the science of the animal and plant breeders, most of whom had never accepted Darwin's theory. The transformation of ideas about heredity around 1900 was brought about by the synthesis of a new social ideology with scientific developments that for the first time focused academic biologists' attention on the work of the breeders. Artificial selection provided the eugenists with the model they needed—but unlike Darwin in the 1830s, they did not see this as a model that could be applied to natural evolution. The new science of heredity—and the social program it sustained—grew out of the non-Darwinian theory of evolution by sudden saltations, and only slowly did it transform in a way that would render it compatible with the theory of natural selection. In a world without Darwin, the selection theory would emerge as an unintended byproduct of the new ideas about heredity prompted by eugenics, thus inverting the causal link presumed by those who see eugenics as a form of social Darwinism.

The Agents of Death

Critics of Darwinism argue that the Nazis' willingness to exterminate rather than sterilize the unfit represents a reemergence of the logic of natural selection. If death is a creative process and individual human life is of little consequence, it makes perfect sense to eliminate those who threaten the biological health of the population. Darwinism also saw humans as little more than highly developed animals, thereby undermining the traditional view of the sanctity of human life. There can be little doubt that Darwin's theory played a role in the destruction of traditional moral and religious values that has characterized modern life. Evolutionism in any form challenged the concept of human uniqueness, although one could mitigate the consequences of this by supposing that the progress of life was somehow intended to produce morally significant results. But natural selection destroyed any hope of seeing purpose in nature: evolution was just trial and error driven by local adaptation and the death of those individuals unfortunate enough to be born with unfit characters. To the extent that people began to see nature in those terms, Darwinism could become a justification for ruthless actions against those who do not contribute to national efficiency.

But were there no other sources of such materialistic attitudes? Can we really believe that a scientific theory so inflamed opinions that without it the world would not have witnessed the horrors of global warfare and genocide? Apart from the huge social changes brought about by industrialization and urbanization, there were a whole series of cultural developments (both in science and in other fields) that made it easier to take a more cynical view of human nature. Natural selection certainly provided a useful model and some effective terminology for those who wanted to project a harsher approach to those deemed less fortunate in nature's lottery. But it was by no means the only source of such attitudes, and the alternative sources would have been quite sufficient to supply the ideologues with the ammunition they needed.

The eminent biologist François Jacob noted that eugenics used the model of artificial, not natural, selection—and that animal breeders are pretty ruthless people who know that to perfect a breed, they must allow only their best specimens to reproduce and must cull the others.[42] Darwin quoted the eminent greyhound breeder Lord Rivers's recipe for success: "I breed many and I hang many."[43] When Darwin and Galton grumbled that humanity alone allows the unfit to breed, they were making a point that could easily have been promoted in complete ignorance of the theory of natural selection. Neither conceived that one might actually eliminate the defectives, but that would be the logical implication of borrowing Lord Rivers's methods. One could argue that the role played by death in Darwin's theorizing about nature came not from Malthus but from the influence of the breeders. The ruthlessness of nature was modeled on that of humanity, not the other way around. The later eugenists who called for the sterilization of the unfit would have had the practices of the breeders in mind, and there would be nothing to stop those seeking to justify more extreme measures from pointing to the fate of the rejected animals.

Jacob also noted that eugenists could have deployed quite different scientific models. For example, Nazis derived one fairly common analogy from medicine. Cancers were seen as an alien force invading the body, and the only way of treating them at the time was to cut them out surgically. Eugenic propaganda often drew an analogy between the threat posed by cancer cells to the body and the threat of alien races breeding within the Aryan population.[44] The implication was that racial cancer had to be excised just as thoroughly. The well-known ethologist Konrad Lorenz, who passed through a phase in which he became willing to endorse Nazi values in the early part of his career, used this analogy. For all his recognition of the role played by aggression in animal and human behavior, Lorenz was no Darwinian. The main reason he offered for supporting eugenics was the need to prevent the spread of inferior racial types that would otherwise contaminate the body of Aryan population.[45]

More generally, there was a growing recognition in the mid-nineteenth century that extinction was a normal part of nature's activity. Geologists had established that the inhabitants of each period in the earth's history were swept away and replaced, and Lyell's uniformitarian geology eliminated the hypothetical catastrophes once thought to be responsible. The only other conclusion was that the decline of populations toward extinction was a gradual and natural process, often triggered by the success of a rival species. Darwin was not the only naturalist to point out that the huge reproductive power of many species meant that most individuals must be destroyed. Naturalists were increasingly concerned about human threats to wild species, and there were already known cases in which human activity had driven species to extinction. Some human tribes, unable to compete with white colonists, had met the same fate—the decline of native groups of Americans and Australians was well established, and the original inhabitants of Tasmania had disappeared.

There was thus no shortage of scientific arguments in the late nineteenth and early twentieth century that could have been used to justify policies of imperial conquest and eugenically inspired genocide. Darwinism helped, of course, but if the theory of individual natural selection had not been available, other sources would have provided adequate arguments with apparent scientific credibility—in particular, the theory of progressive, purposeful evolution that would have emerged, perhaps even more effectively, had Darwin not published.

Even in the more wildly counterfactual world in which the theory of evolution itself became sidelined, other scientific developments would have promoted a more materialistic vision of humanity's place in nature. The development of the neurosciences from phrenology onward would have demolished the concept of the soul by apparently confirming that the brain was the organ of the mind and our mental and moral faculties merely the result of the physical operations of the nerves. The medicalization of human life promoted by a better understanding of the human body and its diseases would have had the same effect. In social theory, the possibility of explaining human behavior in the mass by law-like regularities would also have reduced the autonomy of the individual. Humans would have become mere cogs in a statistical machine.

If Darwin's theory was only one of the many scientific innovations that tended to undermine traditional morality, why is it so persistently singled out as the most important? Can it really be the only factor that, if eliminated, would have enabled us to avoid the Great War and the Holocaust? Darwin's impact has been exaggerated because he became a symbol for the idea of progress through struggle, both in his own time and in the twentieth century. His ideas have been interpreted in different ways over the last hundred and fifty years, but the symbol of Darwin has retained its prominence. As a result, we tend to miss the fact that what he meant to his contemporaries and what he meant to the critics who coined the term "social Darwinism" were very different. In the 1860s and 1870s, he was read very much in the context of Spencer's self-help philosophy, and the harsher implications of the selection theory were sidelined. But in the following generation, he was read as the architect of a worldview based on waste and death (to borrow Samuel Butler's words), and it is this last image that has come down to us through its use by modern creationists. As a result, we have been encouraged to think that the theory of natural selection has permeated the whole evolutionary movement and has had a malign influence throughout its history.

Counterfactual history suggests that a world without Darwin would have lacked the theory of natural selection in the Victorian period, but it would certainly have had the idea of evolution and the ideology of struggle. It would also have had racism and a eugenics movement, although the notion of hereditary determinism would have had other scientific foundations. It would lack not only the theory of natural selection, but also the unifying symbol provided by Darwin himself—and perhaps a pantheon of evolutionary heroes including Spencer and Haeckel would have provided a less attractive target for later critics of the idea of progress. What the theory of natural selection did provide was a battery of ideas and metaphors that—while seldom understood in that way in Darwin's own time—provided ammunition for a later generation of ideologues to use in their calls for the eliminating of the unfit. By showing that rival scientific theories could also have provided equivalent justifications for intolerant attitudes, I have tried to show that Darwin's theory was but one among several potential inspirations for what we call social Darwinism. But this does not rule out the possibility that it was the very language associated with the selection theory that helped to promote, although perhaps not to create, these ideologies. Darwinism allowed the harshness of the Malthusian vision of a world driven by population pressure to sustain into the twentieth century, where it combined with other ideologies to create a nightmare of state-controlled genocide. But was it really the scientific theory that provided the vehicle for this transition?

In a critique of the German language debate on social Darwinism, Richard Evens suggests that what Darwinism provided for the Nazis was not so much a scientific theory as a language though which they could express their ideology of intolerance.[46] Literary scholars such as Gillian Beer and George Levine offer examples of how this could work when they analyze the language in which Darwin presented his ideas and trace parallels in the writings of his contemporaries. Their work can be quite frustrating for the historian of ideas looking for direct connections between scientific and nonscientific concepts. They often ignore the theory of natural selection, and some of the literary figures they examine either did not read Darwin or misunderstood his main purpose. What matters, however, are the metaphors, themes, and attitudes expressed, because they reveal significant resonances between the scientific and the nonscientific prose. This allows us to see how what has been called Darwinian language percolated into the culture of the time, even though Darwin's actual theories may not have made the same transition.

Levine sees Darwinian resources in the writings of Oscar Wilde, for instance, although he concedes that Wilde probably did not read the *Origin* and certainly didn't know much about natural selection.[47] He also sees elements of the grotesque and even of the comic in Darwin's writings, again tracing parallels in the work of literary figures' descriptions of nature. Given that Darwin is the focus of analysis, there seems to be an implication that he may have prompted others to write as they did. But surely there can be no suggestion that Darwin caused the grotesque, for instance, to emerge as a theme in late nineteenth-century literature. He may have highlighted certain aspects of that style and even prompted a few specific examples of it, but no one would seriously suggest that the grotesque would be absent from literature of the period had he not written as he did. By using Darwin as a focus of analysis, scholars project an artificially boosted impression of his influence. They leave us with the feeling that we are seeing a Darwinian element penetrating the language, when what the analysis really shows is that Darwin was a participant in a widely used style of writing. It is time to recognize that just because we can analyze Darwin's language using the techniques developed by literary scholars, this does not mean that every hint of the same language appearing in other texts implies the direct influence of his writing—let alone of his theory.

The same point may also hold for most of the social commentators who drew on evolutionary metaphors. Many did not read Darwin at all and got their ideas about him from popular writers who all put their own spins on the theory. If they did read Darwin, they read him in the context of the other evolutionary writers available, most obviously Spencer, Haeckel, and their followers. Darwinian images and metaphors became common currency without a clear understanding of the theory he proposed—and the fact that he was quoted by ideologues for many different purposes suggests that his theory itself was not the driving force. Taking Darwin and his books out of the equation doesn't remove all the other exponents of evolutionism, and their ideas very often reflected the ideology of progress through struggle expressed in nonselectionist terms. In our world, the whole process became identified with Darwinism, and later generations have chosen to focus only on the negative aspects of the link with science. But in a world without Darwin, the rhetoric of progress through struggle would have emerged in only a slightly less hardheaded manner through the influence of other evolutionists.

What we see in the rise of social Darwinism is not the direct influence of the theory of natural selection but a more diffuse process by which Darwin and his contemporaries exploited possibilities within the available language to express themselves. No doubt Darwin's language shaped the way he was perceived, but eliminating his texts from the discourse wouldn't remove every reference to the struggle for existence. There was, after all, a two-way flow between science and the wider culture based not so much on the interchange of ideas as the pervasiveness of language. We tend to assume that language reflects ideas, but the work of the literary scholars suggests that this may not be the case. Darwin drew on a range of cultural resources available to everyone at the time, and so did other supporters of evolutionism, including those who formulated nonselectionist theories. Darwin was only one source of the language of struggle and extermination, yet he gradually became identified with it, becoming a figurehead for a diffuse influence extending far beyond the theory he proposed. He became a symbol for metaphors and attitudes that were certainly embedded in his theory, but that had a much wider currency both then and in later decades. Without Darwin, the language of progress though struggle might have lacked its cutting edge, but it would still have been available. And people who want to do unpleasant things can always find an excuse for doing so; science may offer resources by which to justify such behavior, but it is hardly likely to initiate it.

CONCLUSIONS

My counterfactual history of evolutionism has been based on the claim that Darwin's unique insights allowed him to create a particular version of the theory of evolution that would not have been available to any of his contemporaries. Darwin retains the respect of biologists today because he came closer than other naturalist of the time to the way of thinking that has underpinned modern evolutionism. (If you doubt this, try reading Spencer, Haeckel, or any of the other late nineteenth-century evolutionists.) He gained these insights not because he was a superhuman genius—although he was very perceptive and persistent—but because he had a combination of interests and experiences that was itself unique for the time. Although historians are not supposed to use such commonsense terminology, I think it can fairly be said that Darwin was "ahead of his time." He disseminated insights that would not otherwise have gained currency for another thirty or forty years. In the non-Darwinian world, the theory of natural selection would probably have been formulated in the early twentieth century, which is when it did at last come into its own in our own world.

Without Darwin's revolutionary input, evolutionism would have developed in a much less confrontational manner, preserving some aspects of the traditional vision of a purposefully designed world and adapting that vision to the modern world via the idea of progress and directed (rather than random) variation. Only when growing concerns about the influence of heredity on human nature prompted the rise of what we call genetic determinism would there be a challenge to the evidence for Lamarckism and forms of directed evolution. Eugenics would encourage recognition of the possible model offered by the animal breeders, allowing for the conception of the idea of selection as a mechanism of natural evolution. Its impact would have been softened, however, by the fact that everyone would appreciate how various developmental processes can shape the effects of genetic mutations. What we call evo-devo would be there from the start, allowing selectionism to be grafted onto the elements of the older tradition that were worth preserving.

In our world, Darwin kick-started the final phase in a long-running campaign to establish an evolutionary worldview by proposing a version of the theory that was far more radical than anything his contemporaries could have anticipated. They were increasingly willing to accept the general idea of evolution, but they found natural selection hard to understand and even harder to accept. Darwin himself probably did not fully appreciate just how damaging his theory was to the idea of a world governed by purpose, since even he seems to have wanted to retain a vestige of the idea of progress. But taken at face value, natural selection depicted the world as the product of trial and error, governed by a relentless struggle for survival. Darwin's religious opponents realized how threatening this was to their traditional faith and reacted accordingly. They dismissed his theory as rank materialism and threw their weight behind the attempt to retain cosmic teleology via the claim that evolution was tending toward a morally significant goal. Liberal religious thinkers were only too willing to jump aboard the progressionist bandwagon. Most of Darwin's contemporaries ignored or subverted the logic of natural selection by proposing that the real source of variation was a purposeful response of an organism to its environment. Selection was accepted only as a negative factor that would eliminate the less successful results of this process.

In a world without Darwin, the concepts underpinning what we call non-Darwinian theories would be the main driving force of a renewed effort to promote evolutionism in the late nineteenth century. Even in our own world, these ideas played a much bigger role than most of us appreciate; without the distraction of Darwinism, they would be the only show in town. There would be no viable concept of completely non-purposeful evolution for materialists to latch onto, and the tradition that runs from T. H. Huxley and John Tyndall to Richard Dawkins and Daniel Dennett might be distracted by the need to seek alternative sources of scientific justification. If the theory of natural selection was not understood primarily as a process of genetic Russian roulette, its value as an argument against design would be muted. By linking natural selection to Malthus and the claim that adaptation is the only factor governing the appearance of new characteristics, Darwin presented his contemporaries with the harshest possible vision of nature. His theory has never been able to shake that image, even though modern ideas about the role of embryological development render it less plausible. In a world without Darwin, both evolutionism and the selection theory would have emerged in a less confrontational manner that would make them much easier for religious thinkers and moralists to accept. Even fundamentalists might see them as just one threat among many, rather than as their chief scientific bogeyman.

And yet the liberals' optimistic vision of social progress collapsed in the decades around 1900, ushering in the era of modernism, with all its doubts about our ability to control or even understand the world. Darwin's religious opponents had predicted just such a collapse, and their modern descendants portray the history of the twentieth century as a vindication of that prediction. The decades around 1900 are seen as the age of social Darwinism during which the analogy of natural selection provided support for the horrors of war and genocide. Liberal religious thinkers were discredited by their alliance with progressionist evolutionism, and their influence was supplanted by neo-orthodoxy and fundamentalism—when it was not eliminated altogether by totalitarian movements of the Right or the Left. Darwinism, if understood in its harshest form, might have predicted this collapse—in a sense it was only offering a different explanation of the flaws in human nature that conservative Christians had always taken for granted.

But how big a role did Darwinism actually play in the collapse of moral values? It certainly cannot be ruled out as a contributor, but our counterfactual analysis has suggested that its critics exaggerate its influence. Without Darwin and his theory, Spencer would not have coined the iconic phrase "the survival of the fittest," and our culture might have lacked the most obvious expressions of the idea that eliminating the unfit was necessary for progress. Spencer's philosophy tended to obscure the harshest interpretation of the phrase, but it remained available for later generations to use in a more horrific manner. Without Darwin, the idea that death plays a constructive role in progress might have taken longer to become established. But we have seen that there were other potential sources from which Nazi ideology could derive its message of hatred and death. If Darwinism were not there to serve as a focus, these other sources would have served almost as well.

Paradoxically, it was Darwin's contemporary opponents who perceived the deeper implications of the theory, while its ostensible supporters did their best to ignore or subvert its more radical implications. Even so, the counterfactual approach allows us to see that most of the effects that have been labeled as "social Darwinism" could have emerged in a world that had no inkling of the theory of natural selection. Some of those effects, most notably scientific racism, might well have been even more strident in the absence of the Darwinian theory. The idea of progress and the allied theories of directed evolution contained the seeds of their own destruction, whether in the imperialists' dismissal of the lower races as relics of the past or in Freud's realization that the higher levels of the mind might not be able to control the primitive foundations still preserved in the unconscious. These and a host of other social and cultural factors all pushed the world toward disaster. Many Christians now say, in effect, that this is just what one could have predicted. But the culprit was not Darwinian pessimism—it was the overoptimistic liberal vision of human perfectibility. Darwin's influence was, at best, to sound an early warning of the potential problems, not to create them.

Deleting Darwin from history would leave us with a science that we could still recognize, although its components would seem to have different implications, thanks to having been assembled in a more natural sequence of discovery. There would be less tension between science and religion, since one of the major battles in what we see as the war between them would never have been fought. But there would have been no deflection of social history onto a track that did not include imperialism, world wars, and episodes of genocide. Scientific ideas do not have that degree of influence, especially when even scientists take decades to appreciate their significance. Evolutionism itself does seem to be implicated in the developments that undermined faith in the progressive program that gave birth to it. But far from being a consequence of Darwin's intervention, that overoptimistic vision of progress was itself a product of wider social and cultural forces. Darwin certainly rocked the boat, but he did not steer it onto a completely new and dangerous course.

NOTES

CHAPTER ONE

1   Darwin was a very poor sailor, and the *Beagle* did indeed run into heavy weather as she attempted to round Cape Horn at the end of 1832. She was nearly sunk by huge waves on January 13, 1833. Fitzroy was present at the famous confrontation over Darwin's theory at the British Association meeting in 1860, where he denounced his former shipmate's theory as contrary to the Bible.

2   Croce, "'Necessity' in History," in *Philosophy, Poetry History*, 557–61. The 2002 edition of Carr's *What Is History?* includes a useful introduction by Richard Evans. My discussion of historians' responses to counterfactualism owes much to Niall Ferguson's introduction to his edited collection *Virtual History*. See also Hawthorn, *Plausible Worlds*. Other collections of counterfactuals by historians include Cowley's *What If?* and *More What If?*; Roberts's *What Might Have Been*; and Snowman's *If I Had Been*.

3   Sobel, *For Want of a Nail*.

4   Gould, *Wonderful Life*. On the butterfly effect, see Lorenz, *The Essence of Chaos*.

5   Morris, *Life's Solutions*.

6   Both essays are in Squire, *If It Had Happened Otherwise*.

7   Fogel, *Railroads and American Economic Growth*.

8   See Marsdon and Smith, *Engineering Empires*, chap. 3 (on steamships); Gooday, *Domesticating Electricity*.

9   Space forbids a detailed outline of the science wars. For the scientists' response to the postmodernist attack, see Gross and Levitt, *Higher Superstition*. The edition cited has an introduction commenting on Alan Sokal's much-publicized hoax article offering a meaningless commentary on modern physics.

10 Radick, "Other Histories, Other Biologies."

11 Stanford, *Exceeding Our Grasp*.

12 Henry, "Ideology, Inevitability, and the Scientific Revolution," a contribution to a section in the journal *Isis* devoted to counterfactualism in the history of science. This includes my own brief outline of the thesis presented in this book, "What Darwin Disturbed," and a valuable introduction by Greg Radick, "Why

What If?" Two other papers offer thoughtful contributions: French's "Genuine Possibilities of the Scientific Past" and Fuller's "The Normative Turn." For other discussions of the possibility of contingency in the history of science, see Arabatzis, "Causes and Contingencies in the History of Science"; and Hacking, *The Social Construction of What?*, esp. chap. 3.

13 The Lamarckian theory assumes that changes to an animal's body brought about by new habits can be transmitted to the offspring. The idea is explored in more detail in several of the chapters that follow.

14 Waller, "Evolution's Inside Man," a contribution to a 2005 *New Scientist* feature on counterfactuals.

CHAPTER TWO

1 Darwin, *Autobiography*, 123–24.
2 Eiseley, *Darwin's Century*, chap. 5; Greene, *The Death of Adam*, 246; and Himmelfarb, *Darwin and the Darwinian Revolution*, chap. 10. The claims that Darwin was not the real discoverer of natural selection are examined in detail later in this chapter.
3 Radick, "Is the Theory of Natural Selection Independent of Its History?"
4 Huxley, "On the Reception of the 'Origin of Species,'" 179–84. For Darwin's quotation, see his *Autobiography*, 141; see also Mayr, *One Long Argument*.
5 On the claim for scientific inevitability, see, for instance, De Beer, *Charles Darwin*; and Mayr, *One Long Argument*. Ideological critics include Himmelfarb, *Darwin and the Darwinian Revolution*; from a Marxist perspective, Robert Young, *Darwin's Metaphor*; and, for the creationists, Weikart, *From Darwin to Hitler*. Note how the modern critics from the Right shift the definition of social Darwinism from cutthroat capitalism to racism and militarism. These issues are discussed more fully in chapter 8.
6 Ruse, *Monad to Man*.
7 In addition to works already cited, accounts of Darwin's life and work include my own *Charles Darwin: The Man and His Influence*; the two-volume biography by Janet Browne; and the joint biography by Adrian Desmond and James Moore, which situates him very firmly in the social debates of the time.
8 All now available in print, see Darwin, *Charles Darwin's Notebooks*; and, for the 1842 and 1844 pieces, Darwin and Wallace, *Evolution by Natural Selection*. The letters reprinted in Darwin, *The Correspondence of Charles Darwin*, also throw much light on his theorizing.
9 Mayr, "Darwin and Natural Selection."
10 See Corsi, *The Age of Lamarck*; Hodge, "Lamarck's Science of Living Bodies"; and Jordanova, *Lamarck*.
11 See, for instance, the various theories described in Rehbock, *The Philosophical Naturalists*.

12 Hodge, "The Universal Gestation of Nature"; Secord, *Victorian Sensation.*

13 Rupke, "Darwin's Choice," offers a preliminary statement of his views.

14 For instance, Richards, *The Meaning of Evolution.*

15 Rehbock, *The Philosophical Naturalists*, chap. 3.

16 Desmond and Moore, *Darwin's Sacred Cause.*

17 Mayr, *One Long Argument.*

18 See Bowler, *The Eclipse of Darwinism, The Non-Darwinian Revolution*, and *Life's Splendid Drama.* These theories are discussed at length in chapters 4 and 5.

19 E.g., Macbeth, *Darwin Retried.*

20 Mayr, *The Growth of Biological Thought* and *One Long Argument.*

21 See Herbert, *Charles Darwin, Geologist.*

22 The most powerful expression of this view is in Desmond and Moore, *Darwin.*

23 Russell, *Form and Function*, offers the standard account of this dichotomy.

24 Ron Amundson's *The Changing Role of the Embryo in Evolutionary Thought* argues the case for a reassessment of the history of evolutionism that will recognize the positive role played by the developmental perspective, as do Nicolaas Rupke's revisionist studies of German structuralism.

25 Kohn, *The Darwinian Heritage*, contains a number of valuable papers on this topic.

26 See Young, *Darwin's Metaphor.*

27 See Bowler, "Malthus, Darwin and the Concept of Struggle"; Herbert, "Darwin, Malthus and Selection"; and Young, *Darwin's Metaphor.*

28 The papers by Wells, Blyth, Matthew, and Wallace are collected in McKinney, *Lamarck to Darwin.*

29 See Eiseley, "Charles Darwin, Edward Blyth, and the Theory of Natural Selection." See also Wainwright, "Natural Selection."

30 Quoted from McKinney, *Lamarck to Darwin*, 49.

31 Ibid., 38.

32 Dempster, *Natural Selection and Patrick Matthew.*

33 See Brooks, *Just before the Origin*; George, *Biologist-Philosopher*; Raby, *Alfred Russel Wallace*; and Williams-Ellis, *Darwin's Moon.*

34 Brackman, *A Delicate Arrangement.* The charge has been repeated more recently by Roy Davies in *The Darwin Conspiracy.*

35 Van Wyhe and Rookmaaker, "A New Theory to Explain the Receipt of Wallace's Ternate Essay by Darwin in 1858," traces the postal deliveries much more carefully and shows that Wallace's paper left the island of Ternate one month later than the date widely assumed.

36 Ruse, "Darwinian Struggles," 417.

37 Wallace to Kingsley, May 7, 1869, Knox Library, Gatesburg. This letter is quoted in James Secord's introduction to Darwin, *Charles Darwin: Evolutionary Writings*, xxi.

38 Osborn, *From the Greeks to Darwin*, 346–48; Poulton, *Charles Darwin*, 80–81; Nicholson, "The Role of Population Dynamics in Natural Selection"; and

Bowler, "Alfred Russel Wallace's Concepts of Variation." For a discussion of the relationship, see Kottler, "Charles Darwin and Alfred Russel Wallace."

39 Quoted from McKinney, *Lamarck to Darwin*, 94.

40 Browne, *Charles Darwin: The Power of Place*, 18; Ruse, *Monad to Man*, 194. See also Smith and Beccaloni, *Natural Selection and Beyond*.

41 Quoted in Poulton, *Charles Darwin*, 79–81.

42 Wallace, *Contributions*, 34.

43 Wallace, *Darwinism*, 64.

44 For a discussion of these differences, see Fichman, *An Elusive Victorian*.

45 Wallace, *Darwinism*, 36–40.

CHAPTER THREE

1 See Van Wyhe, "Mind the Gap." For an account of Darwin's life that stresses the emotional tensions created by his work, see Desmond and Moore, *Darwin*.

2 See Desmond, *Archetypes and Ancestors,* 59. For Huxley's comments, see page 231 in his "Coming of Age of the 'Origin of Species,'" in his *Darwiniana*, 227–43; and his "On the Reception of the 'Origin of Species,'" 197. Haeckel's reminiscences are in his *Last Words on Evolution*, 29.

3 In my *Non-Darwinian Revolution* (48), I used the metaphor of a logjam with Darwin as the lumberjack who freed the logs so the current could push them onward. I now think the jam was not so rigid and that it could have been freed by other, less drastic agents.

4 For detailed accounts of these developments in geology, see Rudwick, *Bursting the Limits of Time* and *Worlds before Adam*.

5 See Bowler, *Fossils and Progress*; Desmond, *Archetypes and Ancestors*; and Rudwick, *The Meaning of Fossils*. On Owen's views, see Rupke, *Richard Owen*.

6 On Erasmus Darwin, see, for instance, Harrison, "Erasmus Darwin's Views on Evolution." More generally on early evolutionism, see Bowler, *Evolution*, chap. 3.

7 Lamarck's theory is outlined in Hodge, "Lamarck's Science of Living Bodies." On the impact of his ideas, see Corsi, *The Age of Lamarck*; on Grant and the British debate, see Desmond, *The Politics of Evolution*. The latter also deals with Owen's response, as does Rupke, *Richard Owen*.

8 On Geoffroy's ideas, see Le Guyader, *Geoffroy Saint-Hilaire*; on his controversy with the more conservative Georges Cuvier, see Appel, *The Cuvier-Geoffroy Debate*.

9 Owen's claims are detailed in Rupke, *Richard Owen*, chap. 5.

10 See Bowler, *Evolution*, chap. 3; and Bowler, *The Mendelian Revolution*. On Mendel's position, see the second edition of Olby's *The Origins of Mendelism*, appendix 5. Lotsy's book is *The Origin of Species by Means of Hybridization*.

11 Roe, *Matter, Life, and Generation*; and Roe, "John Turberville Needham and the Generation of Living Organisms."

12 See Rupke, "Neither Creation nor Evolution" and "Darwin's Choice"; and Gliboff, "Evolution, Revolution and Reform in Vienna."

13 Lyell's letter is in Lyell, ed., *The Life, Letters and Journals of Sir Charles Lyell*, 1:467. On developments in this period, see Bowler, *Evolution*, chap. 4; and Ruse, *The Darwinian Revolution*, chaps. 3–6.

14 There is a modern edition of *Vestiges* edited by James Secord. The best account of Chambers's very non-Darwinian vision of development is Hodge's "The Universal Gestation of Nature."

15 Schaffer, "The Nebular Hypothesis and the Science of Progress"; and Numbers, *Creation by Natural Law*.

16 Secord, *Victorian Sensation*, which includes the responses by Disraeli and Tennyson (see 188–90, 530–32).

17 Rupke, *Richard Owen*, esp. chap. 5

18 See Secord, *Victorian Sensation*, 506–7.

19 Spencer, *Principles of Psychology*, 577. Spencer's 1851 essay "The Development Hypothesis" is reprinted in his *Essays*, 1:381–87.

20 On Baden Powell, see Corsi, *Science and Religion*; for a revisionist account of Spencer, see Francis, *Herbert Spencer and the Invention of Modern Life*. For an overview of the period, see Bowler, *Evolution*, chap.4.

21 See Gliboff, *H. G. Bronn, Ernst Haeckel, and the Origins of German Darwinism*. On Haeckel, see Richards, *The Tragic Sense of Life*; and Di Gregorio, *From Here to Eternity*.

22 Lyell, *Sir Charles Lyell's Scientific Journals*, 84.

23 Francis, *Herbert Spencer*.

24 For a survey of changing attitudes toward the idea of struggle in nature and society, see Gale, "Darwin and the Concept of the Struggle for Existence."

25 Tennyson, *In Memoriam*, stanza 56.

CHAPTER FOUR

1 Beer, *Darwin's Plots*; Levine, *Darwin the Writer*.

2 Huxley, "The Origin of Species," in Huxley, *Darwiniana*, 22–79, see esp. 48.

3 Browne, *Charles Darwin: The Power of Place*, chap. 10.

4 See Robert E. Stebbins's chapter on France in Glick's *Comparative Reception of Darwinism*. See also Bowler, *The Eclipse of Darwinism*, chap. 5.

5 Moore, "Herbert Spencer's Henchmen." See also Moore, *The Post-Darwinian Controversies*; and Bowler, *Monkey Trials and Gorilla Sermons*. These issues are explored in more detail in chapter 7.

6 See Taylor, *The Philosophy of Herbert Spencer*. Mark Francis's *Herbert Spencer and the Invention of Modern Life* argues that Spencer became increasingly pessimistic during the 1860s, although his popular image continued derive from his earlier work.

7   Ruse, *Monad to Man.*

8   For the revolution in archaeology, see Grayson, *The Establishment of Human Antiquity*; Van Riper, *Men among the Mammoths*; and Bowler, *The Invention of Progress.*

9   Stocking, *Victorian Anthropology*; and Kuper, *The Invention of Primitive Society.*

10 The account of this issue in Rupke, *Richard Owen*, chap. 6, presents Owen in a more favorable light than most other studies.

11 Desmond, *Huxley: The Devil's Disciple*, chaps. 13 and 17.

12 For a detailed guide, see Bowler, *Evolution*, chap. 6; and for a revisionary account, see Bowler, *The Non-Darwinian Revolution.* More generally, see Vorzimmer, *Charles Darwin*; Ruse, *The Darwinian Revolution*; Ruse, *Monad to Man*; Kohn, *The Darwinian Heritage*; and Glick, *The Comparative Reception of Darwinism.*

13 The best accounts are Browne, *Charles Darwin: The Power of Place*, chap. 3; and Desmond and Moore, *Darwin*, chap. 33.

14 Haeckel, *Last Words on Evolution*, 29; Haeckel to Darwin, July 9, 1864, in Darwin, *Correspondence of Charles Darwin*, 12:265–59 (trans. 482–85).

15 William Montgomery, "Germany," lists fourteen supporters and six non-Darwinian evolutionists. Other studies, including Sander Gliboff, *H. G. Bronn, Ernst Haeckel, and the Origins of German Darwinism*, identify several names not on Montgomery's list.

16 Hull, Tessner, and Diamond, "Planck's Principle."

17 Montgomery, "Germany," 88.

18 The best modern account of Wallace's work is Martin Fichman, *An Elusive Victorian.* James Moore is conducting an extensive new study, and this section owes much to his advice.

19 Fichman, *An Elusive Victorian*, 80, referring to an 1856 paper on the habits of the orangutan. See also several of the contributions to Smith and Beccaloni, *Natural Selection and Beyond.*

20 Reprinted in Wallace, *Contributions to the Theory of Natural Selection*, see esp. 173 and 200.

21 Bates, "Contributions to an Insect Fauna of the Amazon Valley."

22 On the influence of Wallace's global biogeography, see Bowler, *Life's Splendid Drama*, chap. 8.

23 On the X Club, see Barton, "Huxley, Lubbock, and Half a Dozen Others."

24 On these points, see Bartholomew, "Lyell and Evolution."

25 Lyell, *Sir Charles Lyell's Scientific Notebooks*, 3. In our world, Lyell already knew at this point that Darwin doubted the fixity of species, although he had not yet been told about natural selection.

26 For the quoted phrase, see ibid., 289.

27 Endersby, *Imperial Nature.*

28 See his letter of December 20, 1959, in Darwin, *Correspondence of Charles Darwin*, 7:437.

29 Bellon, "Inspiration in the Harness of Daily Labor." Darwin's book on the fertilization of orchids was published in 1862.

30 On Gray's response to Darwin, see A. Dupree, *Asa Gray*, chap. 13; and on Darwin, Gray, and slavery, see Desmond and Moore, *Darwin's Sacred Cause*.

31 Gray's papers supporting Darwin and reflecting on the theological implications of the theory were collected in his *Darwiniana*; on Wallace's law, see 119 and 191.

32 Ibid., 148.

33 Bartholomew, "Huxley's Defence of Darwinism." See also Di Gregorio, *T. H. Huxley's Place in Natural Science*; and Desmond's two-volume biography, *Huxley: The Devil's Disciple* and *Huxley: Evolution's High Priest*.

34 Huxley, "The Origin of Species," in *Darwiniana*, 22–79, see esp. 77; Huxley, "Mr. Darwin's Critics," in *Darwiniana*, 120–86, see esp. 181–82; Huxley, "Evolution in Biology," in *Darwiniana*, 187–226, see esp. 223.

35 This of course presupposes that Haeckel would write the *Generelle Morphologie* without Darwin's influence, a topic discussed later in this chapter.

36 See the next chapter and, for further details, Bowler, *Life's Splendid Drama*.

37 Owen, "Darwin on the Origin of Species," 496. On Owen's evolutionism, see Rupke, *Richard Owen*, chap. 5.

38 Owen, *On the Anatomy of the Vertebrates*, 3:808.

39 Hull, "Darwinism as a Historical Entity."

40 Herschel's comment is recorded in a letter from Darwin to Lyell, December 10, 1859, in Darwin, *Correspondence of Charles Darwin*, 7:423.

41 I used the term "pseudo-Darwinism" in my *Non-Darwinian Revolution* to denote those biologists who accepted the label "Darwinism" but did not in fact make much use of what we regard as Darwin's most important insights.

42 Müller's book was translated as *Facts and Arguments for Darwin*. See West, *Fritz Muller*.

43 Gliboff, *H. G. Bronn, Ernst Haeckel, and the Origins of German Darwinism*; Richards, *The Tragic Sense of Life*. More sympathetic to my position on Haeckel is Di Gregorio, *From Here to Eternity*.

CHAPTER FIVE

1 The deathbed metaphor was actually used at the time; see Eberhart Dennert's book translated in 1904 as *At the Deathbed of Darwinism*. For the eclipse metaphor, see Huxley, *Evolution*, 22–28. As guides to the vast literature produced by anti-Darwinian evolutionists in our own world, I recommend my own *Eclipse of Darwinism* and *The Non-Darwinian Revolution*.

2 See Eiseley, *Darwin's Century*, chap. 8; Mayr, *The Growth of Biological Thought*.

3 Rupke, "Darwin's Choices"; Reif, "Evolutionary Theory in German Paleontology."

4 Ruse, *Monad to Man*, e.g., chap. 6.

5 Amundson, *The Changing Role of the Embryo in Evolutionary Thought*. See also Laublichler and Maienschein, *From Embryology to Evo-Devo*.

6 On these developments, see Bowler, *Life's Splendid Drama*; and Ruse, *Monad to Man*. In the real world, Müller's book was translated as *Facts and Arguments for Darwin*. Haeckel's popular surveys of evolutionism were translated as *The History of Creation* and *The Evolution of Man*.

7 On these developments, see Bowler, *Fossils and Progress*, chap. 6; Bowler, *Life's Splendid Drama*, esp. chap. 7; Desmond, *Archetypes and Ancestors*; Rudwick, *The Meaning of Fossils*, chap. 5; and Buffetaut, *A Short History of Vertebrate Palaeontology*.

8 See Di Gregorio, *T. H. Huxley's Place in Natural Science*.

9 Broom, *The Mammal-Like Reptiles of South Africa* and *The Coming of Man*.

10 Huxley, "Lectures on Evolution," in his *American Addresses*, 85–90.

11 See Bowler, *Life's Splendid Drama*, chap. 8.

12 Ibid., chap. 9.

13 Ruse, *Monad to Man*, charts the influence of Spencer on a number of key scientists.

14 On Haeckel's vision of nature, see Di Gregorio, *From Here to Eternity*, and Richards, *The Tragic Sense of Life*.

15 Kipling doesn't actually use the example of the giraffe, but "The Elephant's Child" offers a typical example of the effects of stretching, in this case to lengthen the elephant's trunk; see *Just So Stories for Little Children*, 63–84.

16 On the different varieties of Lamarckism, see Bowler, *The Eclipse of Darwinism*, chaps. 4, 5, and 6. For a survey of Lamarckism that connects its nineteenth-century manifestations with modern concerns, see Gisses and Jablonka, *Transformations of Lamarckism*.

17 See, for instance, Mayr, "Germany," in Mayr and Provine, *The Evolutionary Synthesis*, 278–84.

18 See Burkhardt, *Patterns of Behavior*, 338–40. Burkhardt also notes that Alistair Hardy, who became professor of zoology at Oxford, was sympathetic to the quasi-Lamarckian "Baldwin effect."

19 The classic account of Butler's attack is Willey, *Darwin and Butler*; see also Pauly, "Samuel Butler and His Darwinian Critics."

20 Butler, "The Deadlock in Darwinism," reprinted in his *Essays on Life, Art and Science*, see esp. 308

21 On orthogenesis, see Bowler, *The Eclipse of Darwinism*, chaps. 6 and 7.

22 On this extension of the recapitulation theory, see Gould, *Ontogeny and Phylogeny*, esp. chap. 4.

23 Lankester, unpublished "Notes on Variation," quoted in Lester, *E. Ray Lankester*, 89.

24 Radick, *The Simian Tongue*, 368–74. More generally, see Bowler, *The Eclipse of Darwinism*, chap. 8.

25 Galton, *Natural Inheritance*, 27.
26 See Bowler, *The Mendelian Revolution*. The key texts in our world are Bateson, *Materials for the Study of Variation*; De Vries, *The Mutation Theory*; and Morgan's *Evolution and Adaptation* (the latter was written before Morgan was converted to Mendelism; see Allen, *Thomas Hunt Morgan*).

CHAPTER SIX

1 See Cowles, "The Extinction of Alfred Newton."
2 These developments are explored in my *Life's Splendid Drama*, where I argue that they indirectly paved the way for the revival of Darwinism.
3 Osborn's *Origin and Evolution of Life* of 1917 is a far more "modern" book than anything produced in the age of Huxley and Haeckel.
4 See Allen, *Life Science in the Twentieth Century*.
5 The phrase is borrowed from the title of Arthur Koestler's classic but deeply pro-Lamarckian account of the affair, *The Case of the Midwife Toad*. For details, see this book and Bowler, *The Eclipse of Darwinism*, 92–101.
6 See, for instance, Harwood, *Styles of Scientific Thought*.
7 This view was expressed in Bateson's *Materials for the Study of Variation*.
8 On the rise of hereditarian thinking, see Bowler, *The Mendelian Revolution*; Müller-Wille and Rheeinberger, *Heredity Produced*; and Waller, *Breeding*. On how these developments were related to Darwinism, see Gayon, *Darwinism's Struggle for Survival*.
9 Churchill, "August Weismann."
10 Loren Eiseley's *Darwin's Century* is a classic expression of this view.
11 See Dronamraju, *Haldane, Mayr, and Beanbag Genetics*. J. B. S. Haldane was an unashamed supporter of beanbag genetics, Ernst Mayr an opponent—although he did not anticipate the even greater challenge now represented by evo-devo.
12 Harwood's *Styles of Scientific Thought* has already been cited on German genetics; on France, see Burian, Gayon, and Zallen, "The Singular Fate of Genetics in the History of French Biology."
13 For reappraisals of Mendel's work, see Olby, "Mendel No Mendelian"; and Callender, "Gregor Mendel." These and other studies are summarized in Bowler, *The Mendelian Revolution*, chap. 5.
14 Radick, "Other Histories, Other Biologies." Michael Ruse's *Monad to Man* also deals with these biologists in detail.
15 A similar situation emerged after the Second World War, when many German biologists were only too anxious to identify with the dominant paradigm of the English-speaking world. The link between Darwinism and militarism is discussed in more detail in chapter 8.
16 See Bowler, *The Eclipse of Darwinism*, 189, for details. On biometry, see Gayon, *Darwinism's Struggle for Survival*; and Ruse, *Monad to Man*, chap. 6.

17 On the creation of population genetics and the subsequent synthesis with Darwinism, see Provine, *The Origins of Theoretical Population Genetics*; Mayr and Provine, *The Evolutionary Synthesis*; and Smocovitis, *Unifying Biology*.

18 Gould, "The Hardening of the Modern Synthesis."

19 For a comprehensive survey of evo-devo, see, for instance, Carroll, *Endless Forms Most Beautiful*.

20 See Amundson, *The Changing Role of the Embryo*.

21 Pennisi, "The Case of the Midwife Toad." More generally on the reemergence of Lamarckian themes in modern biology, see Gissis and Jablonka, *Transformations of Lamarckism*.

22 Ecology emerged as a distinct science at the end of the century, with inputs from both Darwinian and anti-Darwinian biologists; for details see Bowler, *The Fontana History of the Environmental Sciences*.

CHAPTER SEVEN

1 Other religions, especially Islam, also have problems with evolutionism—although Hinduism, for instance, seems to take the evolutionary perspective on board quite easily. The theory was certainly debated in non-Western cultures from the late nineteenth century onward, but to propose a counterfactual history on such a global scale is beyond my resources.

2 See Turner, "The Victorian Conflict between Science and Religion."

3 See Ruse, *Can a Darwinian Be a Christian?* For detailed accounts of the historical debates, see Moore, *The Post-Darwinian Controversies*; Durant, *Darwinism and Divinity*; and Numbers, *Darwin Comes to America*. I have also provided a survey in my *Monkey Trials and Gorilla Sermons*.

4 He, of course, was "on the side of the angels"; see Moneypenny and Buckle, *The Life of Benjamin Disraeli*, 108.

5 On how scientists were still working to influence public attitudes in the twentieth century, see Clark, *God—or Gorilla*.

6 There were many disagreements on the exact course of human evolution and on how the fossils fit in; see my *Theories of Human Evolution* for details.

7 See Jensen, "Return to the Huxley-Wilberforce Debate"; Lucas, "Wilberforce and Huxley"; and James, "An 'Open Clash between Science and the Church'?"

8 Fichman, *An Elusive Victorian*, chap. 4.

9 Bowler, *Theories of Human Evolution*, chap. 5.

10 Bowler, "Darwinism and the Argument from Design"; for a critique, see Amundson, "Richard Owen and Animal Form," xxxv–xliii.

11 Bowler, "Philosophy, Instinct, Intuition."

12 Osborn, *Man Rises to Parnassus*. On the less optimistic aspects of orthogenesis, see my "Holding Your Head Up High."

13 Broom, *The Coming of Man.*

14 On the literary resonances of Darwinism, see Beer, *Darwin's Plots.* For modern expressions of the positive environmentalist credentials of Darwinism, see Wilson, *The Diversity of Life*; and Levine, *Darwin Loves You.*

15 Gray, *Darwiniana*, 148.

16 Moore, "Herbert Spencer's Henchmen," and more generally his *Post-Darwinian Controversies.* On Spencer's thought in general, see Francis, *Herbert Spencer and the Invention of Modern Life*; and Taylor, *The Philosophy of Herbert Spencer.*

17 The classic expression of this view is Turner's *Between Science and Religion.*

18 Although the British rationalists took up Haeckel's philosophy, monism could also be treated as a form of idealism with religious implications; see my "Monism in Britain" and other articles in Weir, *Monism.* Monism even became associated with esoteric systems such as theosophy.

19 For more details of this transformation, see my "The Specter of Darwinism" and more generally my *Reconciling Science and Religion.*

20 Shaw, *Back to Methuselah*, liv.

21 Koestler, *The Case of the Midwife Toad.*

22 Turner's "The Late Victorian Conflict of Science and Religion" offers a perceptive account of these issues.

23 See Numbers, *The Creationists.* On the Scopes trial, see Larson, *Summer for the Gods.*

24 The classic 1960 movie starring Spencer Tracey was adapted from the 1955 play by Jerome Lawrence and Robert E. Lee. The title *Inherit the Wind* is from Proverbs 11:29.

25 See Livingstone, *Darwin's Forgotten Defenders.*

26 On the local circumstances in Dayton, see Shapiro, "The Scopes Trial beyond Science and Religion."

27 Sulloway, *Freud.*

28 This is "The Voyage That Shook the World," issued by Creation Ministries International in 2009. Although centered on the voyage of the *Beagle*, the program presents the whole evolutionary worldview as an outgrowth of uniformitarian geology. I should add that along with several other historians of science, I allowed myself to be filmed for the program, the makers having concealed the fact that they were backed by a creationist organization.

CHAPTER EIGHT

1 Bryan's position is described well by Gould, *Rocks of Ages*, 155–70.

2 Weikart, *From Darwin to Hitler*; and Weikart, *Hitler's Ethic.*

3 Ruse, *Monad to Man.*

4 Weikart, "Was Darwin or Spencer the Father of Laissez-Faire Social Darwinism."

5 Hawkins, *Social Darwinism in European and American Thought*, 30–35.

6 See Young, *Mind, Brain and Adaptation in the Nineteenth Century*.

7 Huxley's lecture was published in his *Evolution and Ethics*; on his real target, see, for instance, Desmond, *Huxley: Evolution's High Priest*, chap. 10.

8 As Robert Young insists, "Darwinism *is* social"; see his article under that title and more generally his *Darwin's Metaphor*. But that does not mean that it is the only possible scientific expression of the individualist ideology.

9 This point is stressed by Dixon in his *The Invention of Altruism*.

10 See Milam, *Looking for a Few Good Males*; also Richards, "Darwin and the Descent of Woman." For a somewhat uncritical look at how the theory seems to be reflected in the fiction of the time, see Bender, *The Descent of Love*.

11 In addition to Young, *Darwin's Metaphor*, Adrian Desmond and James Moore's *Darwin* stresses his involvement with the social debates of the time. Diane Paul's "Darwin, Social Darwinism and Eugenics" also offers a detailed survey of how his thinking related to later movements. Richard Weikart concedes that both Darwin and Spencer played a role; see his "Was Darwin or Spencer the Father of Laissez-Faire Social Darwinism?"

12 Darwin to Lyell, May 4, 1860, in Darwin, *Correspondence of Charles Darwin*, 8:188–89.

13 Dixon, *The Invention of Altruism*.

14 Hofstadter, *Social Darwinism in American Thought*, chap. 2.

15 Robert Richards's *Darwin and the Emergence of Evolutionary Theories of Mind and Behavior* stresses the moral character of Spencer's philosophy, as does Thomas Dixon in *The Invention of Altruism*.

16 Spencer, *Social Statics*, 378–79. On Spencer, see Francis, *Herbert Spencer and the Invention of Modern Life*; and Taylor, *The Philosophy of Herbert Spencer*. On his wider influence, see Jones and Peel, *Herbert Spencer*.

17 Rylance, *Victorian Psychology and British Culture*, 225.

18 Bannister, *Social Darwinism*. Greta Jones also reminds us that a wide range of political figures, including socialists, used Darwinian rhetoric; see her *Social Darwinism in English Thought*.

19 Quoted by Hofstadter, *Social Darwinism in American Thought*, 45, from Ghent, *Our Benevolent Feudalism*, 29. Ghent notes that the quotation is from a Sunday school address, and he also stresses the magnates' constant appeals to the value of hard work and application.

20 Crook, *Darwinism, War and History*.

21 Darwin, *Descent of Man*, 158.

22 Pearson, *National Life from the Standpoint of Science*. On eugenics, see the final section of this chapter, and on Pearson's early career, see Porter, *Karl Pearson*.

23 See Crook, *Darwinism, War and History*, chap. 3; Hawkins, *Social Darwinism*, chap. 8; and more generally on German Darwinism, Kelly, *The Descent of Darwinism*.

24 Bernhardi, *Germany and the Next War*, 10.

25 Ibid., chap. 4. I have commented on the role played by the idea of successive phases of historical advance in my *Invention of Progress*, chap. 2.

26 See Weindling, *Darwinism and Social Darwinism in Imperial Germany*.

27 On Nietzsche's relationship with Darwinism, see Johnson, *Nietzsche's Anti-Darwinism*; and Moore, *Nietzsche, Biology and Metaphor*.

28 Tennyson, *In Memoriam,* sect. 56; Tennyson wrote the poem between 1833 and 1850.

29 Cowles, "The Extinction of Alfred Newton"; and on Germany, see Nyhart, *Modern Nature*, 116–17.

30 For details, see also Bowler, *Life's Splendid Drama*, chap. 9.

31 Sollas, *Ancient Hunters*, 383. For details of Sollas's and Keith's views, see Bowler, *Theories of Human Evolution*.

32 Sollas, *Ancient Hunters*, 405.

33 See Gould, *The Mismeasure of Man*. For other accounts of race science, see Stepan, *The Idea of Race in Science*; and Haller, *Outcasts from Evolution*.

34 See, for instance, Stocking, *Victorian Anthropology*.

35 See Livingstone, *Adam's Ancestors*.

36 Desmond and Moore, *Darwin's Sacred Cause*.

37 For details, see Bowler, *Theories of Human Evolution*; and Barkan, *The Retreat of Scientific Racism*.

38 Gasman, *The Scientific Origins of National Socialism*. For a response, see Richards, *The Tragic Sense of Life*, appendix 2; also Kelly, *The Descent of Darwinism*. Haeckel would certainly not have imported anti-Semitism from Darwin himself.

39 The most general account is Kevles, *In the Name of Eugenics*; see also Mackenzie, *Statistics in Britain*; and Haller, *Eugenics: Hereditarian Attitudes in American Thought*.

40 On developments in thinking about heredity, see chap. 6; also Waller, *Breeding*; and Müller-Wille and Rheinberger, *Heredity Produced*.

41 Magnello, "The Non-Correlation between Biometrics and Eugenics."

42 Jacob, *Of Flies, Mice, and Men*, 118–19.

43 Quoted by Darwin, *The Variation of Animals and Plants under Domestication*, 2:221.

44 See Proctor, *The Nazi War on Cancer*, 46–47 and chap. 3.

45 See Burkhardt, *Patterns of Behavior*, 244–48.

46 Evans, "In Search of German Social Darwinism."

47 Levine, *Darwin the Writer*, chap.5. The classic study is Beer, *Darwin's Plots*.

BIBLIOGRAPHY

Allen, Garland E. *Life Science in the Twentieth Century*. New York: Wiley, 1975.

———. *Thomas Hunt Morgan: The Man and His Science*. Princeton, NJ: Princeton University Press, 1978.

Amundson, Ron. *The Changing Role of the Embryo in Evolutionary Thought: The Roots of Evo-Devo*. Cambridge: Cambridge University Press, 2005.

Appel, Toby A. *The Cuvier-Geoffroy Debate: French Biology in the Decades before Darwin*. Oxford: Oxford University Press, 1987.

Arabatzis, Theodore. "Causes and Contingencies in the History of Science." *Centaurus* 50 (2008): 32–36.

Bannister, Robert C. *Social Darwinism: Science and Myth in Anglo-American Social Thought*. Philadelphia: Temple University Press, 1979.

Barkan, Elazar. *The Retreat of Scientific Racism: Changing Concepts of Race in Britain and the United States between the World Wars*. Cambridge: Cambridge University Press, 1992.

Bartholomew, Michael. "Huxley's Defence of Darwinism." *Annals of Science* 32 (1975): 525–35.

———. "Lyell and Evolution: An Account of Lyell's Response to the Prospect of an Evolutionary Ancestry for Man." *British Journal for the History of Science* 6 (1973): 261–303.

Barton, Ruth. "Huxley, Lubbock, and Half a Dozen Others: Professionals and Gentlemen in the Formation of the X Club, 1851–1874." *Isis* 89 (1998): 410–44.

Bates, Henry Walter. "Contributions to an Insect Fauna of the Amazon Valley: Lepidoptera: Heliconidae." *Transaction of the Linnaean Society of London* 23 (1862): 495–515.

Bateson, William. *Materials for the Study of Variation: Treated with Especial*

Regard to Discontinuity in the Formation of Species. London: Macmillan, 1894.

Beer, Gillian. *Darwin's Plots: Evolutionary Narrative in Darwin, George Eliot and Nineteenth-Century Fiction.* London: Routledge and Kegan Paul, 1983.

Bellon, Richard. "Inspiration in the Harness of Daily Labor: Darwin, Botany, and the Triumph of Evolution." *Isis* 102 (2011): 393–420.

Bender, Bert. *The Descent of Love: Darwin and the Theory of Sexual Selection in American Fiction, 1871–1926.* Philadelphia: University of Pennsylvania Press, 1996.

Bernhardi, Friedrich von. *Germany and the Next War.* Translated by A. H. Powles. London: E. Arnold, 1912.

Bowler, Peter J. "Alfred Russel Wallace's Concepts of Variation." *Journal of the History of Medicine* 31 (1977): 17–29.

———. *Charles Darwin: The Man and His Influence.* Cambridge: Cambridge University Press, 1996. First published 1990 by Basil Blackwell.

———. "Darwinism and the Argument from Design: Suggestions for a Reevaluation." *Journal of the History of Biology* 10 (1977): 29–43.

———. *The Eclipse of Darwinism: Anti-Darwinian Evolution Theories in the Decades around 1900.* Baltimore: Johns Hopkins University Press, 1983.

———. *Evolution: The History of an Idea.* 1984. Twenty-fifth-anniversary edition. Berkeley: University of California Press, 2009.

———. *Fossils and Progress: Paleontology and the Idea of Progressive Evolution in the Nineteenth Century.* New York: Science History Publications, 1976.

———. *The Fontana History of the Environmental Sciences.* London: Fontana, 1992.

———. "Holding Your Head Up High: Degeneration and Orthogenesis in Theories of Human Evolution." In *History, Humanity and Evolution: Essays for John C. Greene,* edited by James R. Moore, 329–53. Cambridge: Cambridge University Press, 1989.

———. *The Invention of Progress: The Victorians and the Past.* Oxford: Basil Blackwell, 1989.

———. *Life's Splendid Drama: Evolutionary Biology and the Reconstruction of Life's Ancestry, 1860–1940.* Chicago: University of Chicago Press, 1996.

———. "Malthus, Darwin and the Concept of Struggle." *Journal of the History of Ideas* 37 (1976): 631–50.

———. *The Mendelian Revolution: The Emergence of Hereditarian Concepts in Modern Science and Society*. Baltimore: Johns Hopkins University Press, 1989.

———. "Monism in Britain: Biologists and the Rationalist Press Association" In *The Monist Century*, edited by Todd Weir, 175–95. Basingstoke, UK: Palgrave Macmillan, 2012.

———. *Monkey Trials and Gorilla Sermons: Evolution and Christianity from Darwin to Intelligent Design*. Cambridge, MA: Harvard University Press, 2007.

———. *The Non-Darwinian Revolution: Reinterpreting a Historical Myth*. Baltimore: Johns Hopkins University Press, 1988.

———. "Philosophy, Instinct, Intuition: What Motivates a Scientist in Search of a Theory?" *Biology and Philosophy* 15 (2000): 93–101.

———. *Reconciling Science and Religion: The Debate in Early Twentieth-Century Britain*. Chicago: University of Chicago Press, 2001.

———. "The Specter of Darwinism: The Popularization of Darwinism in Early Twentieth-Century Britain," In *Darwinian Heresies*, edited by Abigail Lustig, Robert J. Richards, and Michael Ruse, 48–68. Cambridge: Cambridge University Press, 2004.

———. *Theories of Human Evolution: A Century of Debate, 1844–1944*. Baltimore: Johns Hopkins University Press, 1986.

———. "What Darwin Disturbed: The Biology That Might Have Been." *Isis* 99 (2008): 560–67.

Brackman, Arnold C. *A Delicate Arrangement: The Strange Case of Charles Darwin and Alfred Russel Wallace*. New York: Times Books, 1980.

Bronn, Heinrich Georg. "Essaie d'une réponse a la question de prix propose en 1850 par l'Académie des Sciences . . . savoir: Etudier les lois de la distribution des corps organizes fossils." Supplement, *Comptes rendus de l'Académie des Sciences* 2 (1861): 377–918.

Brooks, John Langdon. *Just before the Origin: Alfred Russel Wallace's Theory of Evolution*. New York: Columbia University Press, 1983.

Broom, Robert. *The Coming of Man: Was It Accident or Design?* London: H. F. and G. Witherby, 1932.

———. *The Mammal-Like Reptiles of South Africa*. London: H. F. and G. Witherby, 1932.

Browne, Janet. *Charles Darwin: The Power of Place*. London: Jonathan Cape, 2002.

———. *Charles Darwin: Voyaging*. London: Jonathan Cape, 1995.

Buffetaut, Eric. *A Short History of Vertebrate Paleontology.* Beckenham, UK: Croom Helm, 1986.

Burian, Richard, Jean Gayon, and Doris Zallen. "The Singular Fate of Genetics in the History of French Biology." *Journal of the History of Biology* 21 (1988): 357–402.

Burkhardt, Richard W., Jr. *Patterns of Behavior: Konrad Lorenz, Niko Tinbergen, and the Founding of Ethology.* Chicago: University of Chicago Press, 2005.

Butler, Samuel. *Essays on Life, Art and Science.* Port Washington, NY: Kennikat Press, 1970. First published 1908 in London.

———. *Evolution Old and New.* London: Harwick and Bogue, 1879.

Callender, L. A. "Gregor Mendel: An Opponent of Descent with Modification." *History of Science* 26 (1988): 41–75.

Carr, E. H. *What Is History?* 1961. Introduction by Richard J. Evans. Basingstoke, UK: Palgrave Macmillan, 2002.

Carroll, Sean B. *Endless Forms Most Beautiful: The New Science of Evo Devo and the Making of the Animal Kingdom.* New York: Norton, 2005.

Chambers, Robert. *Vestiges of the Natural History of Creation and Other Evolutionary Writings.* Edited by James Secord. Chicago: University of Chicago Press, 1994.

Churchill, Frederick B. "August Weismann: A Developmental Evolutionist." In *August Weismann: Selected Letters and Documents*, edited by F. B. Churchill and Helmut Risler, 749–98. Freiburg i. Br.: Universitätsbibliothek, 1999.

Clarke, Constance Areson. *God—or Gorilla: Images of Evolution in the Jazz Age.* Baltimore: Johns Hopkins University Press, 2008.

Cope, Edward Drinker. *Theology of Evolution.* Philadelphia: Arnold, 1887.

Corsi, Pietro. *The Age of Lamarck: Evolutionary Theories in France, 1790–1830.* Berkeley: University of California Press, 1988.

———. *Science and Religion: Baden Powell and the Anglican Debate, 1800–1860.* Cambridge: Cambridge University Press, 1988.

Cowles, Henry. "The Extinction of Alfred Newton: Science and Sentiment in Victorian Animal Protection." *British Journal for the History of Science*, forthcoming.

Cowley, Robert, ed. *More What If? Eminent Historians Imagine What Might Have Been.* London: Macmillan, 2002.

———, ed. *What If? The World's Foremost Military Historians Imagine What Might Have Been.* London: Macmillan, 1999.

Croce, Benedetto. *Philosophy, Poetry, History*. Translated by Cecil Sprigge. London: Oxford University Press, 1966.

Crook, Paul. *Darwinism, War and History: The Debate over the Biology of War from the "Origin of Species" to the First World War*. Cambridge: Cambridge University Press, 1994.

Darwin, Charles. *The Autobiography of Charles Darwin: With the Original Omissions Restored*. Edited by Nora Barlow. New York: Harcourt Brace, 1958.

———. *Charles Darwin: Evolutionary Writings*. Edited by James Secord. Oxford: Oxford University Press, 2008.

———. *Charles Darwin's Notebooks (1837–44)*. Edited by Paul H. Barrett et al. Cambridge: Cambridge University Press, 1987.

———. *The Correspondence of Charles Darwin*. Cambridge: Cambridge University Press, 1985–continuing.

———. *The Descent of Man*. Introduction by Adrian Desmond and James R. Moore. New York: Penguin, 2004.

———. *On the Origin of Species by Means of Natural Selection*. London: John Murray, 1859.

———. *The Variation of Animals and Plants under Domestication*. 2nd ed. 2 vols. London: John Murray, 1882.

Darwin, Charles, and Alfred Russel Wallace. *Evolution by Natural Selection*. Cambridge: Cambridge University Press, 1958.

Davies, Roy. *The Darwin Conspiracy: Origins of a Scientific Crime*. London: Golden Square, 2008.

De Beer, Gavin. *Charles Darwin: Evolution by Natural Selection*. London: Nelson, 1963.

Dempster, W. J. *Evolutionary Concepts in the Nineteenth Century: Natural Selection and Patrick Matthew*. Edinburgh: Pentland Press, 1996.

Dennert, Eberhart. *At the Deathbed of Darwinism*. Translated by E. V. O'Harra and John H. Peschges. Burlington, IA: German Literary Board, 1904.

Desmond, Adrian. *Archetypes and Ancestors: Palaeontology on Victorian London, 1850–1875*. Chicago: University of Chicago Press, 1994. First published 1982 by Blond and Briggs.

———. *Huxley: The Devil's Disciple*. London: Michael Joseph, 1994.

———. *Huxley: Evolution's High Priest*. London: Michael Joseph, 1997.

———. *The Politics of Evolution: Morphology, Medicine, and Reform in Radical London*. Chicago: University of Chicago Press, 1989.

Desmond, Adrian, and James Moore. *Darwin: The Life of a Tormented Evolutionist.* London: Michael Joseph, 1991.

———. *Darwin's Sacred Cause: How a Hatred of Slavery Shaped Darwin's Views on Human Evolution.* London: Allen Lane, 2009.

De Vries, Hugo. *The Mutation Theory: Experiments and Observations on the Origin of Species in the Vegetable Kingdom.* Translated by J. B. Farmer and A. D. Darbyshire. 2 vols. London: Kegan Paul, Trench, Trubner, 1910.

Dick, Philip K. *The Man in the High Castle.* 1962. Introduction by Ben Bova. London: Penguin, 2001.

Di Gregorio, Mario A. *From Here to Eternity: Ernst Haeckel and Scientific Faith.* Goettingen, Germany: Vandenhoek and Ruprecht, 2005.

———. *T. H. Huxley's Place in Natural Science.* New Haven, CT: Yale University Press, 1984.

Dixon, Thomas. *The Invention of Altruism: Making Moral Meanings in Victorian Britain.* Oxford: Oxford University Press, 2008.

Dronamraju, Krishna. *Haldane, Mayr, and Beanbag Genetics.* Oxford: Oxford University Press, 2011.

Dupree, A. Hunter. *Asa Gray.* Cambridge, MA: Harvard University Press, 1959.

Durant, John, ed. *Darwinism and Divinity: Essays on Evolution and Religious Belief.* Oxford: Basil Blackwell, 1985.

Edgerton, David. *The Shock of the Old: Technology and Global History since 1900.* Oxford: Oxford University Press, 2007.

Eiseley, Loren. "Charles Darwin, Edward Blyth, and the Theory of Natural Selection." *Proceedings of the American Philosophical Society* 103 (1959): 94–158.

———. *Darwin's Century: Evolution and the Men Who Discovered It.* New York: Doubleday, 1958.

Endersby, Jim. *Imperial Nature: Joseph Hooker and the Practicalities of Victorian Science.* Chicago: University of Chicago Press, 2008.

Evans, Richard J. "In Search of German Social Darwinism: The History and Historiography of a Concept." In *Medicine and Modernity: Public Health and Medical Care in Nineteenth- and Twentieth-Century Germany*, edited by Manfred Berg and Geoffrey Cocks, 55–79. Cambridge: Cambridge University Press, 1997.

Ferguson, Niall, ed. *Virtual History: Alternatives and Counterfactuals.* London: Picador, 1997.

Fichman, Martin. *An Elusive Victorian: The Evolution of Alfred Russel Wallace.* Chicago: University of Chicago Press, 2004.

Fogel, Robert William. *Railroads and American Economic Growth: Essays in Economic History.* Baltimore: Johns Hopkins Press, 1964.

Francis, Mark. *Herbert Spencer and the Invention of Modern Life.* Stocksfield, UK: Acumen, 2007.

French, Steven. "Genuine Possibilities in the Scientific Past and How to Spot Them." *Isis* 99 (2008): 568–77.

Fuller, Steve. "The Normative Turn: Counterfactuals and a Philosophical History of Science." *Isis* 99 (2008): 576–84.

Gale, Barry C. "Darwin and the Concept of the Struggle for Existence: A Study in the Extra-Scientific Origins of a Scientific Idea." *Isis* 63 (1972): 321–44.

Galton, Francis. *Natural Inheritance.* London: Macmillan, 1889.

Gasman, Daniel. *The Scientific Origins of National Socialism: Social Darwinism in Ernst Haeckel and the Monist League.* New York: American Elsevier, 1971.

Gayon, Jean. *Darwinism's Struggle for Survival: Heredity and the Hypothesis of Natural Selection.* Cambridge: Cambridge University Press, 1998.

George, Wilma. *Biologist-Philosopher: A Study of the Lide and Writings of Alfred Russel Wallace.* New York: Abelard-Schuman, 1964.

Ghent, William J. *Our Benevolent Feudalism.* New York: Macmillan, 1902.

Gibson, William, and Bruce Stering. *The Difference Engine.* 1990. Reprinted London: Vista, 1996.

Gissis, S. B., and E. Jablonka, eds. *Transformations of Lamarckism: From Subtle Fluids to Molecular Biology.* Cambridge, MA: MIT Press, 2011.

Gliboff, Sander, "Evolution, Revolution, and Reform in Vienna: Franz Unger's Ideas on Descent and their Post-1848 Reception." *Journal of the History of Biology* 31 (1998): 179–209.

———. *H. G. Bronn, Ernst Haeckel, and the Origins of German Darwinism.* Cambridge, MA: MIT Press, 2008.

Glick, Thomas F., ed. *The Comparative Reception of Darwinism.* Austin: University of Texas Press, 1974. Reprinted with new preface. Chicago: University of Chicago Press, 1988.

Gooday, Graeme. *Domesticating Electricity: Technology, Uncertainty and Gender, 1880–1914.* London: Pickering and Chatto, 2008.

Gould, Stephen Jay. "The Hardening of the Modern Synthesis." In *Dimensions of Darwinism*, edited by Marjorie Grene, 71–93. Cambridge: Cambridge University Press, 1983.

———. *The Mismeasure of Man.* New York: Norton, 1981.

———. *Ontogeny and Phylogeny.* Cambridge, MA: Harvard University Press, 1977.

———. *Rocks of Ages: Science and Religion in the Fullness of Life.* New York: Ballantine, 1999.

———. *Wonderful Life: The Burgess Shale and the Meaning of History.* London: Hutchinson Radius, 1989.

Gray, Asa. *Darwiniana: Essays and Reviews Pertaining to Darwinism.* New York: Appleton, 1876.

Grayson, Donald K. *The Establishment of Human Antiquity.* New York: Academic Press, 1983.

Greene, John C. *The Death of Adam: Evolution and Its Impact on Western Thought.* Ames: Iowa State University Press, 1959.

Gross, Paul R., and Norman Levitt. *Higher Superstition: The Academic Left and Its Quarrels with Science.* 1994. Reprinted Baltimore: Johns Hopkins University Press, 1998.

Hacking, Ian. *The Social Construction of What?* Cambridge, MA: Harvard University Press, 1999.

Haeckel, Ernst. *The Evolution of Man: A Popular Exposition of the Principal Points of Human Ontogeny and Phylogeny.* New York: Appleton, 1879.

———. *Generelle Morphologie der Organismen.* 2 vols. 1866. Reprinted Berlin: Walter de Gruyter, 1988.

———. *The History of Creation; or, The Development of the Earth and Its Inhabitants by the Action of Natural Causes: A Popular Exposition of the Doctrine of Evolution in General and That of Darwin, Goethe and Lamarck in Particular.* 2 vols. New York: Appleton, 1876.

———. *Kunstformen der Natur.* Leipzig: Bibliographischen Instituts, 1904.

———. *Last Words on Evolution: A Popular Retrospect and Summary.* London: A. Owen, 1906.

Haller, John S. *Outcasts from Evolution: Scientific Attitudes of Racial Inferiority, 1859–1900.* Urbana: University of Illinois Press, 1971.

Haller, Mark H. *Eugenics: Hereditarian Attitudes in American Thought.* New Brunswick, NJ: Rutgers University Press, 1963.

Harrison, James. "Erasmus Darwin's Views on Evolution." *Journal of the History of Ideas* 32 (1972): 247–64.

Harwood, Jonathan. *Styles of Scientific Thought: The German Genetics Community, 1900–1937.* Chicago: University of Chicago Press, 1993.

Hawkins, Mike. *Social Darwinism in European and American Thought, 1860–1945.* Cambridge: Cambridge University Press, 1997.

Hawthorn, Geoffrey. *Plausible Worlds: Possibility and Understanding in the Social Sciences.* Cambridge: Cambridge University Press, 1991.

Henry, John. "Ideology, Inevitability, and the Scientific Revolution." *Isis* 99 (2008): 552–59.

Herbert, Sandra. *Charles Darwin, Geologist.* Ithaca, NY: Cornell University Press, 2005.

———. "Darwin, Malthus, and Selection." *Journal of the History of Biology* 4 (1971): 209–18.

Himmelfarb, Gertrude. *Darwin and the Darwinian Revolution.* New York: Norton, 1959.

Hodge, M. J. S. "Lamarck's Science of Living Bodies." *British Journal of the History of Science* 5 (1971): 323–52.

———. "The Universal Gestation of Nature: Chambers' *Vestiges* and *Explanations.*" *Journal of the History of Biology* 5 (1972): 127–52.

Hofstadter, Richard. *Social Darwinism in American Thought.* Rev. ed. New York: George Braziller, 1959.

Hull, David L., ed. *Darwin and His Critics: The Reception of Darwin's Theory of Evolution by the Scientific Community.* Cambridge, MA: Harvard University Press, 1973.

———. "Darwinism as a Historical Entity: A Historiographical Proposal." In *The Darwinian Heritage*, edited by D. Kohn, 773–812. Princeton, NJ: Princeton University Press, 1975.

Hull, David L., Peter D. Tessner, and Arthur M. Diamond. "Plank's Principle: Do Younger Scientists Accept New Scientific Ideas with Greater Alacrity than Older Scientists?" *Science* 202 (1978): 717–23.

Huxley, Julian S. *Evolution: The Modern Synthesis.* London: Allen and Unwin, 1942.

Huxley, Thomas Henry. *American Addresses.* 1877. Reprinted in volume 4 of *Science and Hebrew Tradition, Collected Essays*, 46–138. London: Macmillan, 1894.

———. *Darwiniana: Collected Essays.* Vol. 2. London: Macmillan, 1894.

———. *Evolution and Ethics: Collected Essays.* Vol. 9. London: Macmillan, 1894.

———. *Man's Place in Nature.* London: Williams and Norgate, 1863.

———. "On the Reception of the 'Origin of Species.'" In *The Life and*

Letters of Charles Darwin, edited by Francis Darwin, 2:179–204. London: John Murray, 1882.

Jacob, François. *Of Flies, Mice, and Men*. Cambridge, MA: Harvard University Press, 1998.

James, Frank A. L. J. "An Open Clash between Science and the Church? Wilberforce, Huxley and Hooker on Darwinism at the British Association, Oxford, 1860." In *Science and Beliefs: From Natural Philosophy to Natural Science*, edited by David Knight and Matthew Eddy, 171–94. Aldershot, UK: Ashgate, 2005.

Jensen, J. Vernon. "Return to the Wilberforce-Huxley Debate." *British Journal for the History of Science* 21 (1988): 161–79.

Johnson, Dirk R. *Nietzsche's Anti-Darwinism*. New York: Cambridge University Press, 2010.

Jones, Greta. *Social Darwinism and English Thought*. London: Harvester, 1980.

Jones, Greta, and Robert A. Peel, eds. *Herbert Spencer: The Intellectual Legacy*. London: Galton Institute, 2004.

Jordanova, L. *Lamarck*. Oxford: Oxford University Press, 1984.

Kelly, Alfred. *The Descent of Darwinism: The Popularization of Darwinism in Germany, 1860–1914*. Chapel Hill: University of North Carolina Press, 1981.

Kevles, Daniel. *In the Name of Eugenics: Genetics and the Uses of Human Heredity*. New York: Knopf, 1985.

Kipling, Rudyard. *Just So Stories for Little Children*. London: Macmillan, 1902.

Knox, Robert. *The Races of Men: A Philosophical Enquiry into the Influence of Race over the Destiny of Nations*. 2nd ed. London: Henry Renshaw, 1862.

Koestler, Arthur. *The Case of the Midwife Toad*. London: Hutchinson, 1971.

Kohn, David, ed. *The Darwinian Heritage: A Centennial Retrospect*. Princeton, NJ: Princeton University Press, 1985.

Kottler, Malcolm Jay. "Charles Darwin and Alfred Russel Wallace: Two Decades of Debate over Natural Selection." In *The Darwinian Heritage*, edited by David Kohn, 367–432. Princeton, NJ: Princeton University Press, 1988.

Kuhn, Thomas S. *The Structure of Scientific Revolutions*. 1962. Reprinted Chicago: University of Chicago Press, 1969.

Kuper, Adam. *The Invention of Primitive Society: Transformations of an Illusion*. London: Routledge, 1988.

Lankester, E. Ray. *Degeneration: A Chapter in Darwinism*. London: Macmillan, 1880.

Larsen, Edward J. *Summer for the Gods: The Scopes Trial and America's Continuing Debate over Science and Religion*. New York: Basic Books, 1998.

Laublicher, Manfred D., and Jane Maienschein, eds. *From Embryology to Evo-Devo: A History of Developmental Evolution*. Cambridge, MA: MIT Press, 2007.

Leakey, Louis S. B. *Adam's Ancestors: An Up-to-Date Outline of What Is Known about the Origin of Man*. 3rd ed. London: Methuen, 1934.

Le Guyader, Hervé. *Geoffroy Saint-Hilaire: A Visionary Naturalist*. Translated by Marjorie Grene. Chicago: University of Chicago Press, 2004.

Lester, Joseph. *E. Ray Lankester and the Making of Modern British Biology*. Edited by Peter J. Bowler. Stanford in the Vale, UK: British Society for the History of Science, 1995.

Levine, George. *Darwin Loves You: Natural Selection and the Re-enchantment of the World*. Princeton, NJ: Princeton University Press, 2008.

———. *Darwin the Writer*. Oxford: Oxford University Press, 2011.

Livingstone, David N. *Adam's Ancestors: Race, Religion, and the Politics of Human Origins*. Baltimore: Johns Hopkins University Press, 2008.

———. *Darwin's Forgotten Defenders: The Encounter between Evangelical Theology and Evolutionary Thought*. Grand Rapids, MI: Eerdmans, 1984.

Lorenz, Edward N. *The Essence of Chaos*. London: University College London Press, 1995.

Lotsy, J. P. *The Origin of Species by Means of Hybridization*. The Hague: Martinus Nijhoff, 1916.

Lucas, J. R. "Wilberforce and Huxley: A Legendary Encounter." *Historical Journal* 22 (1979): 313–30.

Lyell, Charles. *The Life, Letters, and Journals of Sir Charles Lyell*. 2 vols. London: John Murray, 1881.

———. *Sir Charles Lyell's Scientific Journals on the Species Question*. Edited by Leonard G. Wilson. New Haven, CT: Yale University Press, 1970.

Macbeth, Norman. *Darwin Retried: An Appeal to Reason*. Boston: Gambit, 1971.

Mackenzie, Donald. *Statistics in Britain, 1865–1930: The Social Construction of Scientific Knowledge*. Edinburgh: Edinburgh University Press, 1982.

Magnello, Eileeen. "The Non-Correlation between Biometry and Eugenics: Rival Forms of Laboratory Work in Karl Pearson's Career at University College, London." *History of Science* 37 (1999): 79–106, 123–50.

Marsden, Ben, and Crosbie Smith. *Engineering Empires: A Cultural History of Technology in Nineteenth-Century Britain*. Basingstoke, UK: Palgrave Macmillan, 2005.

Mayr, Ernst. "Darwin and Natural Selection." *American Scientist* 65 (1971): 321–27.

———. *The Growth of Biological Thought: Diversity, Evolution, and Inheritance*. Cambridge, MA: Harvard University Press, 1982.

———. *One Long Argument: Charles Darwin and the Genesis of Evolutionary Thought*. Cambridge, MA: Harvard University Press, 1991.

Mayr, Ernst, and William B. Provine, eds. *The Evolutionary Synthesis: Perspectives on the Unification of Biology*. Cambridge, MA: Harvard University Press, 1980.

McKinney, H. Lewis, ed. *Lamarck to Darwin: Contributions to Evolutionary Biology*. Lawrence, KS: Coronado Press, 1971.

Milam, Erika L. *Looking for a Few Good Males: Female Choice in Evolutionary Biology*. Baltimore: Johns Hopkins University Press, 2010.

Montgomery, William M. "Germany." In *The Comparative Reception of Darwinism*, edited by Thomas F. Glick, 81–116. Austin: University of Texas Press.

Monypenny, William F., and George E. Buckle. *The Life of Benjamin Disraeli*. Rev. ed. 2 vols. London: John Murray, 1929.

Moore, Gregory. *Nietzsche, Biology and Metaphor*. Cambridge: Cambridge University Press, 2002.

Moore, James R. "Herbert Spencer's Henchmen: The Evolution of Protestant Liberals in Late Nineteenth-Century America." In *Darwinism and Divinity*, edited by John R. Durant, 76–100. Oxford: Basil Blackwell, 1985.

———, ed. *History, Humanity and Evolution: Essays for John C. Greene*. Cambridge: Cambridge University Press, 1989.

———. *The Post-Darwinian Controversies: A Study of the Protestant Struggle to Come to Terms with Darwin in Britain and America, 1870–1900*. New York: Cambridge University Press, 1979.

Moore, Ward. *Bring the Jubilee*. London: Heinemann, 1955.

Morgan, Thomas Hunt. *Evolution and Adaptation*. New York: Macmillan, 1908.

Morris, Simon Conway. *Life's Solutions: Inevitable Humans in a Lonely Universe*. Cambridge: Cambridge University Press, 2003.

Müller, Fritz. *Facts and Arguments for Darwin*. London: John Murray, 1869.

Müller-Wille, Staffan, and Hans-Jörg Rheinberger, eds. *Heredity Produced: At the Crossroads of Biology, Politics, and Culture, 1500–1870*. Cambridge, MA: MIT Press, 2007.

Nicholson, A. J. "The Role of Population Dynamics in Natural Selection." In *Evolution after Darwin*, edited by Sol Tax, 2:477–522. Chicago: University of Chicago Press, 1960.

Numbers, Ronald L. *Creation by Natural Law: Laplace's Nebular Hypothesis in American Thought*. Seattle: University of Washington Press, 1977.

———. *The Creationists*. Rev. ed. Cambridge, MA: Harvard University Press, 2006.

Nyhart, Lynn K. *Modern Nature: The Rise of the Biological Perspective in Germany*. Chicago: University of Chicago Press, 2009.

Olby, Robert C. "Mendel No Mendelian." *History of Science* 17 (1979): 53–72.

———. *The Origins of Mendelism*. 2nd ed. Chicago: University of Chicago Press, 1985.

Osborn, Henry Fairfield. *From the Greeks to Darwin*. 2nd ed. New York: Scribners, 1929.

———. *Man Rises to Parnassus: Critical Episodes in the Prehistory of Man*. Princeton, NJ: Princeton University Press, 1928.

———. *The Origin and Evolution of Life*. New York: Scribner, 1917.

Owen, Richard. "Darwin on the Origin of Species." *Edinburgh Review* 111 (1860): 487–532. Reprinted in *Darwin and His Critics: The Reception of Darwin's Theory of Evolution by the Scientific Community*, edited by David L. Hull, 175–215. Cambridge, MA: Harvard University Press, 1973.

———. *On the Anatomy of the Vertebrates*. 3 vols. London: Longmans Green, 1866–69.

———. *On the Nature of Limbs*. Edited by Ron Amundson. Chicago: University of Chicago Press, 2007.

Paul Diane B. "Darwin, Social Darwinism and Eugenics." In *The Cambridge Companion to Darwin*, edited by Jonathan Hodge and Greg Radick, 214–39. Cambridge: Cambridge University Press, 2003.

Pauly, Philip J. "Samuel Butler and His Darwinian Critics." *Victorian Studies* 25 (1982): 161–80.

Pearson, Karl. *National Life from the Standpoint of Science.* London: A. and C. Black, 1901.

Pennisi, Elizabeth. "The Case of the Midwife Toad: Fraud or Epigenetics?" *Science* 325 (September 4, 2009): 1194–95.

Porter, Theodore M. *Karl Pearson: The Scientific Life in a Statistical Age.* Princeton, NJ: Princeton University Press, 2004.

Poulton, Edward Bagnell. *Charles Darwin and the Theory of Natural Selection.* London: Macmillan, 1896.

Pratchett, Terry, Ian Street, and Jack Cohen. *The Science of Diskworld.* Vol. 2, *Darwin's Watch.* London: Ebury Press, 2005.

Proctor, Robert N. *The Nazi War on Cancer.* Princeton, NJ: Princeton University Press, 1999.

Provine, William B. *The Origins of Theoretical Population Genetics.* Chicago: University of Chicago Press, 1971.

Raby, Peter. *Alfred Russel Wallace: A Life.* London: Chatto and Windus, 2001.

Radick, Greg. "Introduction: Why What If?" *Isis* 99 (2008): 547–51.

———. "Other Histories, Other Biologies." In *Philosophy, Biology and Life*, edited by Anthony O'Hear, 21–47. Cambridge: Cambridge University Press, 2005.

———. *The Simian Tongue: The Long Debate about Animal Language.* Chicago: University of Chicago Press, 2007.

———. "What If?" *New Scientist* 187 (August 20, 2005): 34–35.

Rehbock, Philip F. *The Philosophical Naturalists: Themes in Early Nineteenth-Century British Biology.* Madison: University of Wisconsin Press, 1983.

Reif, Wolf-Ernst. "Evolution Theory in German Paleontology." In *Dimensions of Darwinism: Themes and Counterthemes in Twentieth-Century Evolutionary Theory*, edited by Marjorie Grene, 71–93. Cambridge: Cambridge University Press, 1983.

Richards, Eveleen. "Darwin and the Descent of Woman." In *The Wider Domain of Evolutionary Thought*, edited by D. R. Oldroyd and Ian Langham, 57–111. Dordrecht: D. Reidell, 1983.

Richards, Robert J. *Darwin and the Emergence of Evolutionary Theories of Mind and Behavior.* Chicago: University of Chicago Press, 1987.

———. *The Meaning of Evolution: The Morphological Construction and Ideological Reconstruction of Darwin's Theory.* Chicago: University of Chicago Press, 1992.

———. *The Tragic Sense of Life: Ernst Haeckel and the Struggle over Evolutionary Thought*. Chicago: University of Chicago Press, 2008.

Roberts, Andrew, ed. *What Might Have Been: Leading Historians on Twelve "What Ifs" of History*. London: Wiedenfeld and Nicolson, 2004.

Roe, Shirley A. "John Turberville Needham and the Generation of Living Organisms." *Isis* 74 (1983): 159–84.

———. *Matter, Life and Generation: Eighteenth-Century Embryology and the Haller-Wolff Debate*. Cambridge: Cambridge University Press, 1981.

Rudwick, Martin J. S. *Bursting the Limits of Time: The Reconstruction of Geohistory in the Age of Revolution*. Chicago: University of Chicago Press, 2005.

———. *The Meaning of Fossils: Episodes in the History of Palaeontology*. 2nd ed. New York: Science History Publications, 1976.

———. *Worlds Before Adam: The Reconstruction of Geohistory in the Age of Reform*. Chicago: University of Chicago Press, 2008.

Rupke, Nicolaas A. "Darwin's Choice." In *Biology and Ideology from Descartes to Dawkins*, edited by Denis R. Alexander and Ronald Numbers, 139–64. Chicago: University of Chicago Press, 2010.

———. "Neither Creation nor Evolution: The Third Way in Mid-Nineteenth Century Thought about the Origin of Species." *Annals of the History and Philosophy of Biology* 10 (2005): 143–72.

———. *Richard Owen: Victorian Naturalist*. New Haven, CT: Yale University Press, 1994.

Ruse, Michael. *Can a Darwinian Be a Christian? The Relationship between Science and Religion*. Cambridge, MA: Harvard University Press, 2001.

———. *The Darwinian Revolution: Science Red in Tooth and Claw*. 2nd ed. Chicago: University of Chicago Press, 1999.

———. "Darwinian Struggles: But Is There Progress?" *History of Science* 47 (2009): 407–30.

———. *Monad to Man: The Concept of Progress in Evolutionary Biology*. Cambridge, MA: Harvard University Press, 1996.

Russell, E. S. *Form and Function: A Contribution to the History of Animal Morphology*. London: John Murray, 1916.

Rylance, Rick. *Victorian Psychology and British Culture, 1850–1880*. Oxford: Oxford University Press, 2000.

Schaffer, Simon. "The Nebular Hypothesis and the Science of Progress."

In *History, Humanity and Evolution*, edited by James R. Moore, 131–64. Cambridge: Cambridge University Press, 1989.

Secord, James A. *Victorian Sensation: The Extraordinary Publication, Reception, and Secret Authorship of "Vestiges of the Natural History of Creation."* Chicago: University of Chicago Press, 2001.

Shapiro, Adam R. "The Scopes Trial beyond Science and Religion." In *Science and Religion: New Historical Perspectives*, edited by Thomas Dixon, Geoffrey Cantor, and Stephan Pumfrey, 198–220. Cambridge: Cambridge University Press, 2010.

Shaw, George Bernard. *Back to Methuselah: A Metabiological Pentateuch.* London: Constable, 1921.

Smith, Charles H., and George Beccaloni, eds. *Natural Selection and Beyond: The Intellectual Legacy of Alfred Russel Wallace.* Oxford: Oxford University Press, 2008.

Smocovitis, Vassiliki Betty. *Unifying Biology: The Evolutionary Synthesis and Evolutionary Biology.* Princeton, NJ: Princeton University Press, 1996.

Snowman, Daniel, ed. *If I Had Been . . . : Ten Historical Fantasies.* London: Robson Books, 1979.

Sobel, Robert. *For Want of a Nail: If Burgoyne Had Won at Saratoga.* London: Greenhill, 2002.

Sollas, W. J. *Ancient Hunters and Their Modern Representatives.* London: Macmillan, 1911.

Spencer, Herbert. *Essays: Scientific, Political and Speculative.* 3 vols. London: Williams and Norgate, 1883.

———. *First Principles of a New Philosophy.* London: Williams and Norgate, 1862.

———. *Principles of Biology.* 2 vols. London: Williams and Norgate, 1864.

———. *Principles of Psychology.* London: Longman, Bown, Green and Longmans, 1855.

———. *Social Statics; or, the Principles Essential to Human Happiness Specified and the First of Them Developed.* London: J. Chapman, 1851.

Squire, J. C., ed. *If It Had Happened Otherwise.* 1932. London: Sidgwick and Jackson, 1972.

Stanford, Kyle. *Exceeding Our Grasp: History and the Problem of Unconceived Alternatives.* Oxford: Oxford University Press, 2006.

Stebbins, Robert E. "France." In *The Comparative Reception of Darwinism*, edited by Thomas F. Glick, 117–67. Austin: University of Texas Press, 1974.

Stepan, Nancy. *The Idea of Race in Science: Great Britain, 1800–1960*. London: Macmillan, 1982.

Stocking, George W. *Victorian Anthropology*. New York: Free Press, 1987.

Sulloway, Frank J. *Freud, Biologist of the Mind: Beyond the Psychoanalytic Legend*. London: Burnett Books, 1979.

Taylor, Michael. *The Philosophy of Herbert Spencer*. London: Continuum, 2007.

Tennyson, Alfred. *In Memoriam*. Edited by Robert H. Ross. New York: Norton, 1973.

Thompson, E. P. *The Poverty of Theory and Other Essays*. London: Merlin Press, 1978.

Turner, Frank Miller. *Between Science and Religion: The Reaction to Scientific Naturalism in Late Victorian England*. New Haven, CT: Yale University Press, 1974.

———. "The Victorian Conflict between Science and Religion: A Professional Dimension." *Isis* 69 (1978): 356–76.

Van Riper, A. Bowdoin. *Men among the Mammoths: Victorian Science and the Discovery of Human Prehistory*. Chicago: University of Chicago Press, 1993.

Van Wyhe, John. "Mind the Gap: Did Darwin Avoid Publishing His Theory?" *Notes and Records of the Royal Society* 60 (2007): 177–205.

Van Wyhe, John, and Kees Rookmaaker. "A New Theory to Explain the Receipt of Wallace's Ternate Essay by Darwin in 1858." *Biological Journal of the Linnean Society* 105 (2012): 249–52.

Vorzimmer, Peter J. *Charles Darwin: The Years of Controversy: The "Origin of Species" and Its Critics, 1859–82*. Philadelphia: Temple University Press, 1970.

Wainwright, Milton. "Natural Selection: It's Not Darwin's (or Wallace's) Theory." *Saudi Journal of Biological Sciences* 15 (2008): 1–8.

Wallace, Alfred Russel. *Contributions to the Theory of Natural Selection*. 2nd ed. London: Macmillan, 1871.

———. *Darwinism: An Exposition of the Theory of Natural Selection*. London: Macmillan, 1889.

———. *The Geographical Distribution of Animals*. 2 vols. London: Macmillan 1876.

Waller, John. *Breeding: The Human History of Heredity, Race and Sex*. Oxford: Oxford University Press, forthcoming.

———. "Evolution's Inside Man." *New Scientist* 187 (August 20, 2005): 43–44.

West, David A. *Fritz Müller: A Naturalist in Brazil*. Blacksburg, VA: Pocahontas Press, 2003.

Weikart, Richard. *From Darwin to Hitler: Evolutionary Ethics, Eugenics, and Racism in Germany*. New York: Palgrave Macmillan, 2004.

———. *Hitler's Ethic: The Nazi Pursuit of Evolutionary Progress*. New York: Palgrave Macmillan, 2009.

———. "Was Darwin or Spencer the Father of Laissez-Faire Social Darwinism?" *Journal of Economic Behavior* 71 (2009): 20–28.

Weindling, Paul J. *Darwinism and Social Darwinism in Imperial Germany: The Contribution of the Cell Biologist Oscar Hertwig*. New York: Gustav Fischer, 1991.

Weir, Todd, ed. *Monism: Science, Philosophy, Religion, and the History of a Worldview*. Basingstoke, UK: Palgrave Macmillan, 2012.

Willey, Basil. *Darwin and Butler: Two Versions of Evolution*. London: Chatto and Windus, 1960.

Williams-Ellis, Amabel. *Darwin's Moon: A Biography of Alfred Russel Wallace*. London: Blackie, 1966.

Wilson, Edward O. *The Diversity of Life*. Cambridge, MA: Harvard University Press, 1992.

Young, Robert M. "Darwinism *Is* Social." In *The Darwinian Heritage*, edited by D. Kohn, 609–38. Princeton, NJ: Princeton University Press, 1985.

———. *Darwin's Metaphor: Nature's Place in Victorian Culture*. Cambridge: Cambridge University Press, 1985.

———. *Mind, Brain and Adaptation in the Nineteenth Century: Cerebral Localization and Its Biological Context from Gall to Ferrier*. Oxford: Clarendon Press, 1970.

INDEX

adaptation, 143–44, 148–52, 176; and Darwinism, 39, 47–49, 112, 116, 138, 195–200; and design, 39, 48–49, 74, 77, 207; and Lamarckism, 77–78, 138, 157–59, 207; rejection of by biologists, 119, 121–22, 162–67
Agassiz, Louis, 106, 108, 161, 260
agnosticism, 99, 223
Allen, Garland E., 177
alternation of generations, 80
altruism, 243. *See also* morality, evolution of
America: Darwinism in, 95, 117–18, 229–32, 243, 246–47; genetics in, 173, 179, 189, 192–93, 200
Amundson, Ron, 142–43, 201
anatomy. *See* morphology
ancestral inheritance, law of, 183–86, 196
anthropology, 8, 102–3, 257–60
apes, 103–4, 125, 209–14, 231
archaeology, 101–2, 258
Archaeopteryx, 148
argument from design. *See* design
Argyll, Duke of, 218
artificial selection, 50–51, 57, 64, 128, 172, 182–83, 197, 264, 266–71
autochthonous generation, 81. *See also* spontaneous generation

Babbage, Charles, 82, 83–84
Bagehot, Walter, 251
Balfour, Francis, 145
Bannister, Robert, 246
Bartholomew, Michael, 121
Bates, Henry Walter, 112
Bateson, William, 166, 178, 179–80, 191–92, 195
Beagle, voyage of, 1, 36–37, 40–41, 45
Beecher, Henry Ward, 225

Beer, Gillian, 9, 273
Bellon, Richard, 116
Bergson, Henri, 161, 228
Bernhardi, Friedrich von, 252–23
biogeography, 9, 170; evolutionism and, 39–41, 60, 108–9, 110–18, 152–54; social implications of, 254–56
biomedical sciences, 20, 272
biometry, 195–96
Blumenbach, J. F., 39, 48, 78, 141, 257
Blyth, Edward, 44, 55–56
botany, 115–18
breeding, of animals and plants. *See* artificial selection
British Association for the Advancement of Science, 212
Broca, Paul, 257, 260
Bronn, Heinrich Georg, 86, 127–28
Broom, Robert, 214, 218
Browne, Janet, 63, 92
Bryan, William Jennings, 230, 233, 248
Buckle, Henry, 236
Buffon, Georges-Louis Leclerc de, 81
Butler, Samuel, 160–61, 221, 227–28, 239, 273
butterfly effect, 9

Campbell, Reginald, 227
cancer, as analogy for racial degeneration, 271
capitalism: as form of social Darwinism, 242–47; as influence on Darwinism, 32–33, 46–47
Carlyle, Thomas, 7
Carnegie, Andrew, 246–47
Carpenter, William Benjamin, 85, 126, 218
Carr, E. H., 6, 7
catastrophism, in geology, 74
chain of being, 36, 139, 211, 257

Reif, Wolf-Ernst, 142
religion: and human origins, 209–17; Lamarck-
ism and, 209, 221–29; opposed to Darwin-
ism, 25, 207–8, 217–20, 229–32; opposed
to evolution, 24–25, 84–85, 205–6, 229–
31; supportive of evolution, 98, 220–29.
See also design; fundamentalism; theistic
evolution
revolutions in science, 14, 22, 95, 200, 203
Richards, Robert J., 130
Rivers, Lord, 270–71
Rockefeller, John D., 246–47
Rolleston, George, 107, 123
Rookaaker, Kees, 58
Rupke, Nicolaas, 39, 81, 141–42
Ruse, Michael, 32, 59, 63, 131, 142, 234

saltations, 78, 122, 166–68, 182, 191–92, 268–69
Schopenhauer, Arthur, 253
science, objectivity of, 12–14, 17, 32, 197. *See
also* revolutions in science
Scopes, John Thomas, 212, 230
Secord, James, 83, 87, 97
Sedgwick, Adam, 45, 84, 106
sexual selection, 133, 239–40
Shaw, George Bernard, 221, 228–29
slavery, 41, 260
Smith, Adam, 8, 47
Sobel, Robert, 9, 11
social Darwinism, 233–35; Darwin and, 2–3,
24–27, 242–43; definition of, 235–42; non-
Darwinian versions of, 3, 88, 223, 243–47.
See also social evolution
social evolution, 233–35; Darwinism and, 25–
27, 242–43; Lamarckism and, 85, 88, 242–
47. *See also* progress, in social evolution
Sollas, William Johnson, 256, 261
soul, origin of, 209–10, 212, 272
species, nature of, 43–44, 110–12, 116, 167, 182
speech, origin of, 166–67
spontaneous generation, 80–81, 83
Spallanzani, Lazzaro, 80
Spencer, Herbert: on Lamarckism, 222–26,
244–47; and Malthus, 32, 55, 89; on naturalism, 99–100, 209, 221–25; as promoter
of evolution, 19–20, 70, 85–86, 88, 92, 95,
155–56; on self-improvement, 181, 243–
46; on struggle for existence, 160, 209, 239;
on "survival of the fittest"; 43, 88, 243
Stanford, Kyle, 15
Sterling, Bruce, 11

structuralism. *See* formalism
struggle for existence, 55; between nations,
154, 247–56; between species, 170; Darwin as symbol of, 26, 173–74; in Darwin's
theory, 51–53, 62, 204; Lamarckism and,
160, 224–26; within society, 84, 88–90,
223, 237–39. *See also* natural selection
Sumner, William Graham, 246
"survival of the fittest," 43, 100, 243, 246,
278. *See also* natural selection; struggle for
existence

technology, history of, 11–12
teleology. *See* design; theistic evolution
Tennyson, Alfred, 84, 88, 89–90
theistic evolution, 65, 85–86, 118, 160, 218,
220–21
Thompson, E. P., 8
Tolstoy, Leo, 7
tree of life. *See* common descent
Treitschke, Heinrich von, 253
Trevelyan, G. M., 10
Tschermak, Erich von, 177
Tylor, Edward B., 102, 258
Tyndall, John, 215

Unger, Franz, 81
uniformitarianism, 45–46, 74–75, 113, 232

Van Wyhe, John, 58, 67
variation, random, 50–51, 173, 185, 188, 196–
97, 207. *See also* orthogenesis; saltations
varieties (subspecies), 44, 62–63, 110–12,
115–16, 130, 170
vertebrates, origin of, 146–47, 159, 178
Vestiges of the Natural History of Creation, 67–
68, 82–85, 97, 215
Virchow, Rudolf, 107, 134
vitalism, 160–61, 226, 228
Vogt, Karl, 81, 86, 261

Wallace, Alfred Russel: on biogeography, 41,
108–9, 113, 152–54; as co-discoverer of
natural selection, 1, 29, 47, 53–54, 58–66,
73, 110–11, 170–71; on human origins, 212,
260; as promoter of evolution, 21, 31, 51,
59–60, 110
Waller, John, 21
war, 248, 252–53. *See also* Civil War; imperialism; struggle for existence; World War I;
World War II